【江苏省绿色智慧建筑（新一代房屋）】系列丛书
"Jiangsu Province Intelligent Green Buildings (Next Generation Houses)" Book Series

智慧树
垂直社区的未来生活
建筑设计国际竞赛获奖作品集

A Collection of Award-winning Works of the
"Smart Tree — Future Life in a Vertical Community"
International Architectural Design Competition

绿色智慧建筑（新一代房屋）课题组 编
Edited by Intelligent Green Buildings (Next Generation Houses) Project

中国建筑工业出版社

编委会

主　　任：周　岚　　顾小平
副 主 任：刘大威
主　　编：路宏伟　　刘永刚
副 主 编：张　彤　　汪　杰　　程建军　　王登云　　杨小冬　　姚　京
编写人员：丁　杰　　赵双双　　项菲菲
图书设计：王　江　　汪　昊　　陈　岑

EDITORIAL COMMITTEE

Directors: Zhou Lan, Gu Xiaoping

Deputy Director: Liu Dawei

Editors-in-Chief: Lu Hongwei, Liu Yonggang

Deputy Editors-in-Chief: Zhang Tong, Wang Jie, Cheng Jianjun, Wang Dengyun, Yang Xiaodong, Yao Jing

Contributors: Ding Jie, Zhao Shuangshuang, Xiang Feifei

Book Designers: Wang Jiang, Wang Hao, Chen Cen

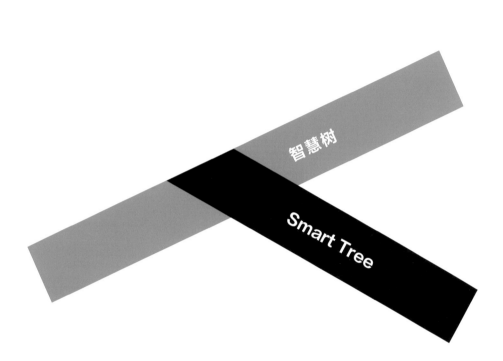

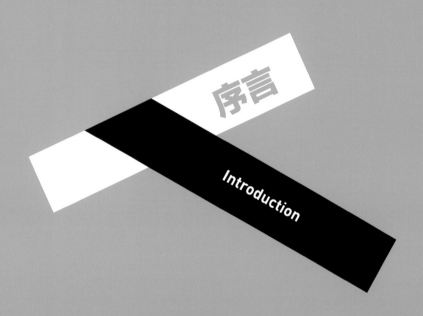

序言一　Introduction 1

走向泛建筑学
Comprehensive Architecture in the Making

孟建民，博士；教授；中国工程院院士；全国工程勘察设计大师；第七届梁思成建筑奖获得者；深圳市建筑设计研究总院有限公司总建筑师。
Meng Jianmin;Ph.D, Professor, Academician at the Chinese Academy of Engineering; National architecture design master; Winner of Liang Sicheng Architecture Prize; Chairman of BOD / Chief Architect of SADI; Principal of Meng Architects of Architectural Society of China.

以下是孟建民院士在 2018 年 4 月 28 日未来绿色智慧建筑国际研讨会上做的题为"泛建筑学"的演讲的部分内容。

The following is part of speech entitled Comprehensive Architecture by Academician Meng Jianmin in the International Symposium of Futuristic Intelligent Green Buildings on April 28, 2018.

纵观 150 亿年宇宙发展史，建筑的发展历程微不足道。100 年、1000 年、甚至 1 万年以后的建筑，还和现今一样吗？在一般理解中，建筑始终是一种静止、凝固、永恒的艺术存在。然而，随着当今科技的迅猛发展，人类的生存和生活状态不断发生实质性的变化，那么未来的建筑学也会因此而被重新定义吗？

In the vast universe which is 15 billion years old, the development of architecture is nearly negligible. Will architecture be the same as today in 100 years, 1000 years or even 10,000 years time? Normally people think architecture is a static, fixed and eternal form of art. Yet as the current era is seeing a rapid development of science and technology as well as essential changes in people's condition of existence and living, will architecture be redefined in the future?

自工业革命以来，科技进步就与建筑及人类生存空间的发展息息相关。一方面，科技进步改变了人类的生存方式，拓展了建筑的类型，且不断刷新建筑高度，保障建筑安全性。另一方面，科技进步也提升了轮船、飞机、空间站等载人工具的制造技术水准，丰富了它们的功能形态，以此延伸了人类的生存空间。

Since the period of the Industrial Revolution, scientific and technological development has been closely related to human's living spaces. On the one hand, scientific and technological progress has changed people's way of living, and resulted in an increase in categories, height and security of buildings. On the other hand, scientific and technological development also means improved functions and standards of manufacturing technologies such as vessels, aircrafts and space stations among other manned systems, thus the living space of humans have been further expanded.

未来的建筑学必将是顺应科技发展，并且结合其他人造物、科技产物而存在的日新月异的学科。如果说传统的建筑学以工程学、风景园林学、城市规划学等关联学科为支撑，那未来的泛建筑学，则把边界广泛延伸至物理学、航天科技、生物学、化学、材料学、人工智能、信息技术等一系列交叉学科。研究"泛建筑学"这一前瞻性概念，正是为了借助全方位的科技力量，将智能化真正应用于人的需求。

Architecture of the future should follow the trend of technological development and integrate accordingly with fast advancing fields related to artificial products and technological products. Traditional architecture is based on relevant disciplines including engineering, landscape architecture and urban planning; while comprehensive architecture in the future would follow an interdisciplinary approach by extending fringes to physics, aerospace technology, biology, chemistry, materials science, artificial intelligence and information technology among others. Research on the forward-looking concept "comprehensive architecture" is intending to apply technologies of various fields to aspects that really meet people's needs.

从"建筑学"走向"泛建筑学"，不仅打破了建筑与其他制造门类的界限，同时也将模糊衣食住行的边界，让建筑的功能与内容融于人类的各种生存形式中。建筑终将成为有机的、生态的、智慧的，甚至是可以呼吸的、具有自我修复能力的生命体。

The development from "architecture" to "comprehensive architecture" means breaking the boundaries of architecture and other fields, as well as obscuring the distinction between the needs for clothing, food, accommodation and traveling, thus integrating functions and elements of architecture to every aspect of people's lives. Buildings are sure to become organic beings that are intelligent, eco-friendly and even breathable and capable of self-restoration.

序言二 Introduction 2

智慧树竞赛的国际发布
International Announcement of the 'Smart Tree' Competition

缪昌文,中国工程院院士,东南大学教授、博导;国务院政府特殊津贴获得者;江苏省建筑科学研究院有限公司董事长;国际绿色建筑联盟主席;国际材料与结构试验研究联合会(RELEM)管理与决策委员会委员;江苏省人民政府参事。

Miao Changwen is an academician of Chinese Academy of Engineering, a professor and PhD supervisor of Southeast University, and a winner of the Governmental Special Allowance of the State Council. He is the chairman of Jiangsu Research Institute of Building Science Co., Ltd., and concurrently serves as the president of International Green Building Alliance, a member of Management and Decision-making Committee of International Union of Laboratories and Experts in Construction Materials, Systems and Structures (RELEM), and a counsellor of People's Government of Jiangsu Province.

国际绿色建筑联盟作为一个国际性交流合作创新平台，旨在推动形成绿色发展方式和生活方式进程中努力实现绿色建筑理念融通、技术联通、标准相通、人才互通，共商共建共享绿色建筑发展的新未来。

The International Green Building Alliance, as an international platform for communication, cooperation and innovation, aims to, in the process of promoting the formation of a green development method and life style, make efforts to realize the conceptual integration, technical connectivity, standard interlinking, talent interflow, and jointly negotiate on, co-construct and share the new future of the green building development.

江苏省绿色建筑发展已有十余年，实现了从试点示范到全面推进，从行业领域的探索到城市范围的拓展和区域的集成。江苏的绿色建筑走出了一条从理念到实践、不断探索创新的发展之路。

In Jiangsu Province, the green building has developed for more than a decade, and realized the transformation from pilot demonstration to overall promotion, and from the exploration in industrial fields to the expansion in urban scope and the integration in the region. The green building in Jiangsu Province has stepped onto a development road of from concept to practice, and of constant exploration and innovation.

2017 年，省委十三届三次会议把推动高质量发展作为当前和今后一个时期的根本要求，提出了城乡建设高质量等六个高质量发展目标。如何回应和落实高质量发展要求，是当下绿色建筑发展应当积极思考和研究的。为此，2018 年江苏开展了绿色智慧建筑（新一代房屋）课题研究，旨在探索江苏未来的建筑如何更好地满足人居生活需求、更加高质量的发展。

In 2017, the Third Meeting of the 13th Session of the Provincial CPC Committee took promoting high-quality development as a fundamental requirement at present and in the future, and brought forward six high-quality development objectives such as urban and rural high-quality construction, etc. How to respond to and implement the high-quality development requirements is what should be actively considered and researched for green building development. Therefore, Jiangsu developed the topic research on Intelligent Green Buildings (Next Generation Houses) in 2018, aiming to explore how the future buildings of Jiangsu could meet human settlement and living needs still better, and realize higher-quality development in the future.

本次建筑设计国际竞赛采用真题实做方式，以江苏未来建筑试点项目作为标的物，吸引了来自境内外十多个国家的高等院校、设计机构近百组选手参赛。孟建民院士、Vicente Guallart 先生、沈迪、李兴钢等 13 位建筑设计大师作为本次的评审专家，共评选出 26 组获奖作品。本书内容包含了各位大师对未来建筑发展的思考和获奖作品介绍，希望以此书的出版为契机，能有更多的专家学者参与进来，共同为江苏省未来建筑高质量发展建言献策。

This International Architectural Design Competition, by means of practical doing of real thing as the exam, and with the pilot project of future building in Jiangsu as subject matter, has attracted nearly 100 groups of participants of universities and colleges, and design institutions from domestic and overseas more than ten countries. Total 13 architectural design masters, including the academician Meng Jianmin, Mr. Vicente Guallart, Shen Di, and Li Xinggang, etc., as review experts of this competition, have selected 26 groups of award-winning works through appraisal. This book contains the thinking of each master about the future building development and the introduction to the award-winning works. It is expected that, via the publication of this book, more experts and scholars could participate in and bring forward suggestions for the high-quality development of future buildings in Jiangsu Province.

目录 Contents

1	**背景介绍** Background Introduction		49	**建筑模块定制篇** Chapter of Architectural Module Customization

1 背景介绍
Background Introduction

13 获奖作品集
Collection of Award-winning Works

15 建筑系统优化篇
Chapter of Architectural Systematic Optimization

16 建筑的自然代谢
BUILDING METABOLISM

24 漂浮的原野——生长的未来居住空间
SUSPENDED FIELDS —— Future Living Space for Growth

28 孟建民：科技拓展建筑的边界
Meng Jianmin: Expand the Boundaries of Architecture via Science and Technology

30 张彤：未来建筑的智慧变革
Zhang Tong: Intelligent Revolution of Buildings of the Future

33 建筑生态创新篇
Chapter of Architectural Ecological Innovation

34 云舟
Cloud Boats

40 拉近——智慧的垂直生活理念
Closer —— Smart Vertical Llife Concept

44 Vicente Guallart：科技重构城市生态
Vicente Guallart: Science and Technologies Restructure Urban Ecosystems

46 Hernan Diaz Alonso：建筑是城市变革的推动者
Hernan Diaz Alonso: Architecture is the Promoter of Urban Change

49 建筑模块定制篇
Chapter of Architectural Module Customization

50 未来垂直村——AI 匹配的装配式自进化栖居
The Future of Vertical Village —— AI Matched Fabricated Self-Evolving Residence

56 WONDERLAND——传统生活的新回归
WONDERLAND —— The New Return of Traditional Living

62 垂直聚落
Vertical Cluster

66 4C 生活：可持续性互联集装箱社区
4C LIFE: The Container Community of Sustainability and Interconnection

70 智慧树——未来垂直生活社区
SMART TREE: Future Life in a Vertical Commuity

74 HUB——垂直智慧社区
HUB —— Vertical Intelligent Community

76 All You Need is BOX
All You Need is BOX

80 定制生活，从盖房子开始……
Customized Life Starts with Building…

84 "拼图式"垂直花园
Jigsaw Vertical Garden

86 李兴钢：垂直社区的生态定制
LI Xinggang: Ecological Customization of Vertical Community

88 张应鹏：高适应性与功能化的未来建筑
Zhang Yingpeng: Architecture in the Future Would be Highly Adaptable and Have Various Functions

90 Michael U. Hensel：超越隐喻，思考未来
Michael U. Hensel: Go Beyond the Metaphor When Reflecting on the Future

93	**智慧社区生活篇**
	Chapter of Smart Community Life

94	"DNA"双螺旋垂直街巷——街区重构
	"DNA" Double Helix Vertical Street-Block Reconstruction
100	定制生活，共享社区
	Private Customized Space & Shared Community
102	宜居 +
	GREEN LIVING +
106	垂直社区的未来生活——青年创客住宅
	Future Life in a Vertical Community—Residential Design for Young Maker
108	UPLIFE/ 共享生活：住宅之外更是空间乃至全身心的改变
	UPLIFE/Shared Life: Beyond Dwelling: Transforming Spaces, and even Minds and Hearts
110	塔罗社区
	TARRO Community
114	城中游牧：移居与一体化生活
	Nomadic City: Integrative Life of Movability
116	沈迪：科技重塑邻里社区
	Shen Di: Technologies Reshape Neighborhoods and Communities
118	冯正功：在延续中守护未来
	Feng Zhenggong: Protecting the Future in Continuation
120	Daniel V. Hayden：体验创造终极未来
	Daniel V. Hayden: Experience Creating the Ultimate Future

123	**建筑科技应用篇**
	Chapter of Architectural Technology Application

124	Verti- Community
	Verti- Community
130	智核绿巢
	Intelligent Nuclear Green Nest
134	智能细胞系统——未来垂直生活社区
	SMARTCELL SYSTEM- Future Life in a Vertical Community
138	ADAPTIVE.COM MUNITY
	ADAPTIVE.COM MUNITY
140	Uplift Microcity
	Uplift Microcity
144	垂直森林
	Vertical Forest
148	徐卫国：构筑数字建筑的产业网链
	Xu Weiguo: Constructing the Industrial Network Chain of Digital Buildings
150	Henriette H. Bier: 机器人建筑时代
	Henriette H. Bier: Architecture in the Age of Robotic Building
152	张雷：生态科技 回归自然
	Zhang Lei: Eco-technology, Returning to the Nature

155	**回顾和鸣谢**
	Review and Acknowledgement

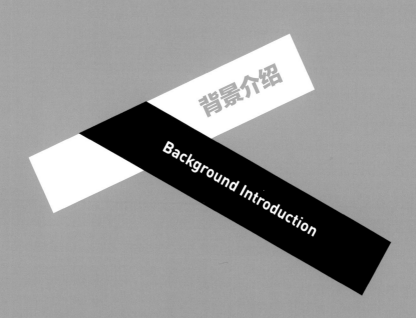

背景介绍一
Background Introduction 1

推动江苏省新一代建筑高质量发展
——江苏省绿色智慧建筑（新一代房屋）研究与实践

Promote High-quality Development of Next Generation Architecture in Jiangsu Province
——Research and Practice of "Jiangsu Province Intelligent Green Buildings (Next Generation Houses)"

江苏省绿色智慧建筑"新一代房屋"课题组
"Jiangsu Province Intelligent Green Buildings (Next Generation Houses)" Project

江苏，自古便是理想人居地的代表，这里孕育了丰富多彩的自然与文化。今天的江苏，既是我国经济社会的先发地区，同时也是人口、资源和环境压力最大的省份之一。江苏城镇化水平高出全国十多个百分点，在快速城镇化推进过程中，资源环境的约束日益趋紧、矛盾日益突出。2008年以来，围绕转型发展和可持续人居家园的目标，江苏抓住快速城镇化时期大量建设的机会，努力推进城乡建设向绿色发展方向转变。十多年来，江苏通过省级建筑节能专项引导资金支持，有效地推进了各地建筑节能、绿色建筑和绿色生态城区的探索和实践，从试点示范到全面推进，从小规模的探索到城市范围的拓展和集成，绿色发展的实践渐次深入，内涵不断完善，内容日益综合。

Jiangsu has been an ideal place to live since ancient times. There are diversified natural sceneries and cultures. Although Jiangsu's economic and social development started earlier than many other provinces, it also faces great pressure from population, resources and environment at present. The urbanization level in Jiangsu is over 10% higher than the national average. Fast urbanization also comes along with increasingly great limitation and noticeable contradiction concerning resources and environment. Since 2008, Jiangsu Province has focused on changing development mode and building sustainable living conditions, and has facilitated the shift to green development mode in enormous town and country construction programs during the period of fast urbanization. Over the ten years, Jiangsu Province has effectively facilitated attempts and practices in energy conservation, eco-friendly buildings and ecological city. From pilot demonstration projects to comprehensive implementation, from small scale practice to expansion and consolidation across the whole city, green development is gaining ground and keeps improving and integrating with various elements.

十九大报告将"坚持人与自然和谐共生"作为新时代中国特色社会主义思想和基本方略，提出推进绿色发展，建设人与自然和谐共生的现代化要求。2019年国务院政府工作报告提出"加强污染防治和生态建设，大力推动绿色发展""促进资源节约集约和循环利用，推广绿色建筑"。国家生态文明和绿色发展的要求，以及江苏的省情特点，都要求江苏必须率先探索绿色发展道路。

In the report Xi Jinping delivered at the 19th National Congress of the Communist Party of China, "harmonious coexistence between man and nature" is regarded as an important concept and basic strategy of socialism with Chinese characteristics in the new era. To fulfill this goal, we also need to promote the green development. The report also includes requirements on enhancing pollution prevention and control, promoting ecological development, vigorously advancing green development" and "boosting resource conservation and recycling and encouraging the construction of green buildings. Considering national requirements on promoting ecological development and green development as well as local conditions of the province, Jiangsu have to take the initiative to pursue green development.

面对新时代绿色发展的新要求，人民群众对居住品质和环境提升的新需求，城乡建设高质量发展的新目标，江苏率先设立"江苏省绿色智慧建筑（新一代房屋）"研究，旨在通过对未来人居生活需求和社会环境变化的研判，打破传统建筑分专业的机械设计思维，以建筑全生命周期为研究对象，通过对绿色、健康、长寿、智慧、人文等关键要素的系统、整合和优化，从建筑设计方法创新、建筑空间可变探索、建造施工安全高效、室内环境健康舒适、科技产品适度运营、智慧人居运营管理等多方面开展深入研究，提升建筑领域科技水平，实现建筑科技水平与人居环境品质双提升。

In response to green development requirements in the new era, people' new demand in living standards and environment upgrading, as well as new targets for high quality of urban and rural development, Jiangsu Province proactively initiated the research program of "Jiangsu Province Intelligent Green Buildings (Next Generation Houses)". By doing so they aim to evaluate living demands and social environment changes in the future and break away from inflexible setting of majors in traditional architecture industry. This research regards life cycle of buildings as subject, and systematizes, consolidates and optimizes key factors related to green, health, longevity, intelligence and humanity among others. They have conducted in-depth research on multiple aspects including innovation in architecture design method, flexible building spaces, safety and efficiency of construction, healthy and comfortable indoor environment, appropriate operation of tech products, operation and management of intelligent living system. By doing so, they have improved living conditions and technological standards of the architecture industry.

"智慧树——南京江北新区国际人才公寓"项目作为"江苏省绿色智慧建筑（新一代房屋）"研究的示范项目之一，课题组拟将其打造成2020～2035年居住建筑的未来模样。2018年，课题组在威尼斯国际建筑双年展上发布国际建筑设计竞赛，收获了来自境内外高等院校、设计机构和个人的参赛作品近百份，凝聚全球智慧为智慧树项目提供了更多发展方向和技术理念。本书是对26个获奖作品的系统总结，同时还梳理了孟建民院士、Vicente Guallart先生、沈迪、李兴钢、张彤、张雷等建筑设计大师对未来建筑的思考。希望以此为起点，进一步推进江苏省新一代房屋的研究与实践，扩大示范引领作用，以点带面，推动江苏绿色建筑高质量发展。

"Smart Tree — International Talent Department in Jiangbei New Area of Nanjing" is one of the demonstration projects of "Jiangsu Province Intelligent Green Buildings (Next Generation Houses)" research program. The research group hopes to build this site with futuristic architecture style of years between 2020 and 2035. In 2018, the research group announced international architectural design competition in Venice Architecture Biennale. They received around 100 entries to the competition from institutions of higher education, design firms, and individuals at home and abroad. Intelligent people across the world provided Smart Tree project with many new strategies on development directors and technological ideas. This book has made a comprehensive summary of all 26 award-winning solutions. Moreover, this book also included some great architects' insights on architecture in the future. Among them are Academician Meng Jianmin, Vicente Guallart, Shen Di, Li Xinggang, Zhang Tong and Zhang Lei. We hope this could become the starting point for more researches and practices concerning Next Generation Houses in Jiangsu Province. Moreover, we hope this project would play a bigger role in demonstration and guidance so as to encourage much more practices, thus promoting high-quality development of green buildings in Jiangsu Province.

背景介绍二
Background Introduction 2

"下一代建筑"发展计划
"The Next Architecture" Project

雅伦格文化艺术基金会
Fondazione EMGdotART

科学技术的迅速发展，以原子能、电子计算机、空间技术和生物工程的发明和应用为主要标志的第四次科技革命，涉及了信息技术、新能源技术、新材料技术、生物技术、空间技术和海洋技术等诸多领域，让人类正经历着一场空前规模的数字化信息技术革命。这场革命不仅极大地推动了人类社会经济、政治、文化领域的变革，为世界文化的发展提供了雄厚的物质基础，且影响了人类生活方式和思维方式。

This era has seen rapid development in science and technology. The fourth scientific and technological revolution mainly features inventions and applications of atomic energy, electronic computer, space technology and bioengineering. This revolution involves information technology, new energy technology, new material technology, biotechnology, space technology and marine technology among others. All these have resulted in an unprecedented great revolution on digital information technologies. This revolution has not only vigorously promoted reform in economy, politics and culture of human society, but also laid a solid foundation for advance in the world civilization, and also influenced human's living and thinking patterns.

在一场由科技、数字信息与创造力引领的全球数字化转型和智慧革命下，建筑作为最有潜力的科技、设计、数据化以及艺术和创意的集成平台，正在数字化大潮中被重新定义，也正迎来前所未有的机遇和挑战。至少我们已经认识到空间和资源的总量是有限的，地球环境和气候正在因为包括建造在内的人类活动经历不可逆转的变化。大数据和云计算基础设施的急速迭代和完善，使得以人工智能和物联网为代表的信息技术迅速改变着我们的生活、思维和社会的运行模式。我们会同时面对身处环境的生态危机，以及来自数据与算法进化的另一种"智慧体"的挑战。

Against the backdrop of global digital transformation and intelligent revolution guided by technology, digital information as well as creative ideas, architecture is being redefined as an integrated platform of technology, design, digitization, art and creativity. Moreover, this sector is also faced with unparalleled challenges and opportunities. At least we have realized that the total amount of space and resources is limited, and that the earth's environment and climate are enduring irreversible changes caused by various human activities including those of construction industry. With ultra-rapid iteration and improvement in the infrastructure of big data and cloud computing, information technologies featuring artificial intelligence and Internet of Things (IoT) are changing the modes of our life, thinking and society in a fast speed. Meanwhile, we would face ecological crisis and challenges from a new kind of "intelligent beings"created due to advances in data analysis and algorithms.

在这样的背景下，作为塑造环境、承载生活的最大量基础设施，建筑会发生什么样的改变？当构建房屋的不再是砖石，而是数据；当建筑既是数据的入口，又是数据沉淀和分享的平台；当数据在人、建筑和应用之间循环累进……建筑是否会被诠释成为一种数据生态？当房屋具备了"生命体"机制，自身不断学习和持续进化，我们是否做好准备要接受它的思考、行动和感情？我们是否可以想象建筑从自发无序的无机体脱离，进化成为自然生态系统的组成部分，并形成以"连接

+大数据智能+科技+自适应服务"为基本范式的更高层级智慧体……

As buildings serve as the greatest infrastructure that shapes environment and provides livelihood support, what change will happen to them? When data plays a larger role than bricks in constructing buildings, when buildings become both the entrance of data and platform for accumulating and sharing data, when data is being optimized during its continuous circulation among people, buildings and applications…, will buildings be regarded as an ecosystem of data? If houses resemble living creatures and can keep learning and evolving by themselves, can we get prepared to accept their ideas, actions and emotions? Can we imagine that buildings are no longer inorganic beings and evolve into an integral part of the natural ecosystem, and even become higher level intelligent beings with basis on norms of "connection+big data+technologies +self-adaptation services" …

在此基础上,雅伦格文化艺术基金会(Fondazione EMGdotART),联合DADA(中国建筑学会数字设计专业委员会)、IAAC(西班牙加泰罗尼亚高等建筑研究院)、城市复兴2050(UR2050)等共同发起「下一代建筑」发展计划,并将"下一代建筑"全球创新大赛作为重要组成部分,以全球智慧连接与共享为基础,聚集那些应对能源和环境压力,通过加载新的科技并利用大数据而不断增强解决问题能力,创造丰富的公共性、流动互联的物联网生活,从资源角度实现环境平衡和低生态足迹,以及开拓先锋可行的智能建造技术等前沿理念、科研、技术和系统化解决方案,形成示范效应。

Against this backdrop, Fondazione EMGdotART cooperated with agencies including China's Digital Architecture Design Association (DADA), Spain's Institute for Advanced Architecture of Catalonia (IAAC), and Urban Regeneration 2050 (UR2050). They jointly initiated "The Next Architecture" Project, and the "Next Architecture" Global Innovation Competition is an integral part of this plan. Based on global mechanism of intelligent connection and sharing, we intend to jointly tackle pressures concerning energy and environment, apply new technologies and use big data to improve problem solving abilities, and thus create a flexible and connected IoT system with diversified public elements. We hope to maintain an environmental balance and low carbon footprint from the perspective of resources, and also explore pioneering and feasible cutting-edge concepts, conduct scientific researches, and work out technological and systemic solutions for intelligent construction and others fields. By doing so, we hope we can set a good example.

"下一代建筑"发展计划及全球创新大赛正是基于对未来的探索,对新的技术、人才、科研、创意等,以建筑为研究对象和应用载体,进行的大规模的聚集、引入与协作,从而带来全新的解决方案、运行模式、不断创新迭代的空间功能产品以及由大数据的应用而派生出的全新服务价值等。

"The Next Architecture" Project and Global Innovation Competition are intended to explore the future of technologies, talents, scientific researches and innovations and other aspects concerning architecture. By aggregating, introducing and coordinating relevant resources, and with architecture as the research object and carrier to be applied to, we hope that there will emerge brand new solutions, operation modes, products with building spaces' functions in continuous innovations and iteration, as well as new service values stemming from application of big data, to name a few.

未来充满呼唤和感召!"下一代建筑"发展计划及全球创新大赛将在更长的时间尺度上引领研发、突破传统、勇于创新,拥抱由大数据和科技发展所带来的深度变革,以新的方式、生态和能力走向未来。

Future never cease to appeal to us! "The Next Architecture" Project and Global Innovation Competition would guide research and development, break with traditions, make innovations and embrace the profound reform brought about by big data and technological development over a very long period of time. Thus we would acquire new skills, build new ecosystems and embrace the future in a new way.

背景介绍三
Background Introduction 3

智慧树：垂直社区的未来生活国际设计竞赛
Smart Tree: International Design Competition for Future Life in Vertical Communities

"智慧树——垂直社区的未来生活"建筑设计国际竞赛组委会
Organizing Committee of the "Smart Tree —— Future Life in a Vertical Community" International Architectural Design CompetitionT

什么是下一代建筑？一方面人类星球正在经历不可预期的环境气候变化，另一方面以人工智能和物联网为核心的数据技术正在迅速改变我们的生活。"垂直智慧社区"体现对亚洲高密度城市中正在发生并可预期的未来生活之理解与回应；同时，项目的研发和建设将促进代表智能化、信息化、清洁能源等新科技产业与建筑设计和社区系统综合管理的多方面融合。

What is 'Next Architecture'? On one hand, our planet is experiencing unpredictable changes in climate change. On the other hand, digital technology cored with artificial intelligence and the Internet of Things has changed our lives dramatically. "Vertical Intelligent Community" embodies an understanding and response to ongoing phenomena and life expectancy in densely populated cities in Asia. At the same time, the research and development as well as construction of this project will advance integration of new technology industries, promoting intelligence, informatization and clean energy along with architectural design and the general management of community systems.

在江苏省"未来屋"课题和"下一代建筑"发展计划的指引下，国际绿色建筑联盟、东南大学、南京长江都市建筑设计股份有限公司、中建八局第三建设有限公司、南京国际健康城开发建设有限公司、雅伦格文化艺术基金会联合主办"智慧树——垂直社区的未来生活"国际设计竞赛，共同探索下一代建筑。

Guided by the "Jiangsu Province Intelligent Green Buildings(New Generation House)"project and "The Next Architecture"project; which intends to explore the next architecture," International Design Competition of Smart Tree: Future Life in Vertical Communities" and is jointly co-organized by the International Green Building Alliance, Southeast University, Nanjing Yangtze River Urban Architectural Design Co., Ltd, The Third Construction Co., Ltd. of the Eighth Bureau of China Construction, Nanjing International Healthcare Development Co., Ltd., and Fondazione EMGdotART.

竞赛标的介绍
Introducing the Competition

竞赛标的基于中国江苏省南京市江北新区国际健康城人才公寓 3 号楼，江北新区国际健康城人才公寓位于江北新区国际健康城——一个以居住功能为主的绿色低碳园区内，其中 3 号楼总建筑面积 22270 平方米，总高度 100 米，预计容纳 450 人居住。将该幢建筑地面层以上 35.1 米高的区段作为竞赛范围（标的）。从地面到 15.3 米为垂直社区公共空间（B 部分），15.3～35.1 米为未来居住概念体验单元（A 部分）。

The competition is based in Building No. 3 of the International Healthcare Area Talent Apartment in Jiangbei New Area, Nanjing, which is located in a green, low-carbon park in a residential area of Jiangsu Province. The total construction area covers 22,270 square meters, with a total height of 100 meters. It is expected to accommodate 450 people in total. 35.1-meter-high section above the ground floor of the apartment building will be utilized as the competition scope. From the ground up to 15.3 meters (Part B) will be designated as the public vertical community space. From 15.3 meters to 35.1 meters (Part A) will be designated as the conceptual experiment unit of the future residential area.

中国江苏省南京市江北新区
Jiangbei New Area, Nanjing, Jiangsu Province, China

江北新区国际健康城基地位置
Location of the International Healthcare Area in Jiangbei New Area

江北新区国际人才公寓 3 号楼
Building No. 3, International Talent Apartment, Jiangbei New Area

功能内容不限于未来居住概念单元，但主要为居住单元，其他内容包括小型研发及头脑风暴中心、创客空间、商务会议、图书馆、展厅、冷链超市、餐厅、咖啡厅、茶室、健身、影院、农场、花园等功能，并能实现最大限度的灵活可变性。

Functional content includes, but is not limited to, conceptual future residential units. However, these are mainly residential units. Other possible content includes small-scaled R&D and brainstorming centers, makerspaces, business meeting rooms, libraries, exhibition halls, cold-chain supermarkets, restaurants, coffee shops, tearooms, fitness clubs, movie theaters, farms, gardens and other functions that can maximize flexibility and variability.

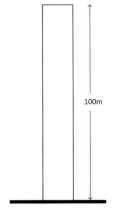

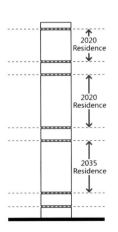

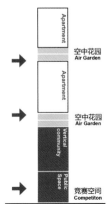

竞赛任务书规定参赛方案要强调整体性、集成化和科技研发的成分，体现：高层建筑的能源效率解决方案（如可再生能源、分布式能源系统以及单元模块能源使用的集成效率等）；以数据驱动设计，形成建筑的智慧大脑；结合人工智能与物联网技术，对人工智能时代的居住、娱乐、研发、交流等空间模式和技术应用场景进行探索与回应；以及智能建造技术等。正如"下一代建筑"发展计划所相信的，建筑的数据生态亦可被诠释为建筑将会变成服务与信息网络的个性化终端，就像手机是网络信息的智能终端一样，建筑亦将是我们生活中网络智能化的终端。不仅如此，以建筑内的人、设备、环境等感知信息作为数据源泉，以物联网技术为支撑实现万物互联，将数据汇聚到一个开放平台，可以开放给创新、创业者进一步开发使用。该生态系统永远在线，以标准服务接口的方式提供给人工智能、虚拟现实服务、支持创新人员以群体智慧的方式开展社会化协同创新，围绕建筑产生可自我进化的多样性应用服务，不断改善用户体验、丰富生活方式，激发并增强个体的创造力，以及参与智慧型社会的能力。

According to requirements on the competition task, contestants' solutions should emphasize wholeness, integration as well as scientific and technological R&D. Their plans should include: solutions for energy efficiency of high-rise buildings (such as renewables, distributed energy system and integrated efficiency of energy module units); establishing an intelligent system of the building with data-driven design; exploring spatial models and technology use cases in living, entertainment, R&D and communication among others in the era of artificial intelligence; and intelligent construction technologies, etc. As described in "The Next Architecture" project, buildings' data ecosystem could mean buildings become personalized terminal of services and information network. As mobile phone is an intelligent terminal of information network, building can also become a terminal of intelligent network in our life. Nonetheless, with people, devices, environment among other perceptual information serving as data sources, with IoT technologies becoming the support for connectivity, data can be converged in one open platform for further development and uses of innovators and entrepreneurs. This ecosystem will be always online and use standard interfaces to provide services based on artificial intelligence and virtual reality technologies, support creative people to use collective wisdom to make innovations featuring social coordination. Thus they would put in place diversified application services centered on buildings and these applications would be able to evolve automatically. And they would improve user experience and enrich people's lives. This would spur creativity of individuals and boost their ability to get involved in a smart society.

竞赛任务介绍
Introducing the Competition Task

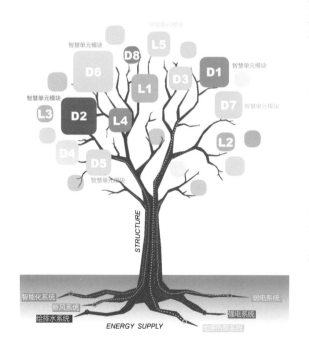

1. 南京市江北新区国际健康城人才公寓"智慧树－垂直社区"竞赛要求设计统筹，科技赋能，向全球建筑师、设计师、科研机构、院校、科技企业、创意先锋、研发团队等征集整体解决方案，包括（或可任选）以下三项内容板块：
1）整体系统组织与整合性设计；
2）多种未来生活单元模块集成（可基于功能设定）；
3）关键性技术和科技创新。
基于竞赛主体即将建造实现，因此设计方案需考虑建造可行性与技术方案。

1. In general, the "Smart Tree-Vertical the International Healthcare Area Talent Apartment in Jiangbei New Area, Nanjing, calls for overall design coordination and enabled technology. It invites global architects, designers, research and development institutions, academies, enterprises, and innovative pioneers to present integral solutions, including (or selected from) the following three content blocks:
1) Integral systematic organization and integrated design;
2) Integration of various future living modules (can be based on functional settings);
3) Key technologies and scientific innovation.
As the competition project will be constructed and realized, the design solution must therefore be reliable and practical, both in construction and technical feasibility within the design scheme.

2. 竞赛鼓励参赛者 / 团队以联合体（设计 + 科研 + 技术）的形式，从"下一代建筑"的以下（不限于）七项要求出发，提出对应下述内容的创新解决方案。七项要求（关键性设计原则）：

1）在高密度垂直生活中创造丰富的公共性；
2）适应最大限度的可变性；
3）流动互联的物联网生活；
4）从资源角度实现环境平衡；
5）开放融合的垂直生态，实现低生态足迹；
6）创造美的环境；
7）先锋而可行的智能建造技术。

3. 参赛者需重点考量：
1）亚洲城市高密度条件下垂直生活的公共性；
2）高层建筑的能源效率解决方案，包括但不限于可再生能源、分布式能源系统以及单元模块能源使用的集成效率等；
3）以数据驱动设计，形成建筑的智慧大脑，不仅带来高效的系统运行，更能从细微之处贴近人的情感和需求，使空间抵达生活的层次与人性的融合；
4）结合人工智能与物联网技术，对人工智能时代的居住、娱乐、研发、交流等空间模式和技术应用场景进行探索与回应；
5）结合"智慧树"竞赛对象所在环境（建筑、园区、城市），配置体现未来生活的居住个性、公共性、交互性、智慧性等方面的功能内容，可容纳但不限于未来居住概念单元、小型研发及头脑风暴中心、创客空间、商务会议、图书馆、展厅、冷链超市、餐厅、咖啡厅、茶室、健身、影院、农场、花园等功能，并能实现最大限度的灵活可变性；
6）在智能建造方面提供具有可实施性的技术方案。

4. 以下为加分项：
1）参赛者以联合团队形式带入科技解决方案，包括智能建造、物联网、信息技术、能源、新材料、人工智能、VR&AR 等；
2）提出由于技术优化和系统协同设计等带来的在建筑运营、（能源）管理等方面的经济效益测算 / 评估，以及（或者）由于功能配置等方面带来的商业创新模式。

2. The competition advocates all participants as individuals or groups (a joint team consisting of members focusing on design + R&D + technology) to provide innovative solutions including, but not limited to, the following 7 requirements of "The Next Architecture" project (key design principles):

1) Create abundant publicity in high-density and vertical life.
2) Adapt to maximum variability.
3) Utilize the Internet of Things, with mobility and connectivity.
4) Achieve environmental balance from a resource perspective.
5) Achieve a low ecological footprint in open and integrated situations.
6) Create a pleasant and aesthetic environment.
7) Pioneer a practical and intelligent construction technology.

3. Key points which need to be considered by participants:
1) The publicity of vertical life under high-density conditions in Asian cities.
2) Energy efficiency solutions for high-rise buildings, including but not limited to the integrated efficiency of renewable energy, distributed energy systems, and energy efficiency in unit modules.
3) Developing an intelligent system of the building with data-driven design, which not only brings efficient systematic operation, but also takes subtle human emotions and needs into account to enable the space by integrating to the level of living and humanity.
4) Combining the artificial intelligence and the Internet of Things to explore and respond to space patterns and technical application scenarios such as residence, entertainment, research and development, communication and other spatial patterns, in this era of artificial intelligence.
5) Combining with the environment (buildings, parks, and cities) where participants of the "Smart Tree" are create configurations that embody the functional content that best reflects residential personality, publicity, interaction, and wisdom of the future life, which can accommodate but are not limited to the conceptual units of future residences, small-scaled R&D and brainstorm centers, makerspaces, business meetings, libraries, exhibition halls, cold-chain supermarkets, restaurants, coffee shops, tearooms, fitness clubs, cinemas, farms, gardens, etc., in which will also achieve maximum flexibility and variability.
6) Provide practical technical solutions in intelligent construction.

4. The following qualify as bonus points:
1) Joint-team participants bring in technology solutions, including intelligent construction, the Internet of Things, information technology, energy, new materials, artificial intelligence, VR & AR, etc.
2) Propose the economic evaluation of economic benefits in terms of building operations, energy management and/or business innovative models originating from functional configurations that come along with technical optimization and systematic collaboration design.

竞赛机制
Competition Mechanism

1. 奖励机制
1) 一等奖共 1 组，每组颁发奖杯、证书和奖金 25000 欧元（含税）；
2) 二等奖共 2 组，每组颁发奖杯、证书和奖金 12000 欧元（含税）；
3) 单元模块奖共 3 组，每组颁发奖杯、证书和奖金 6000 欧元（含税）；
4) 入围奖 20 组，每组颁发证书；
5) 特别奖 1 名，下一代建筑奖学金奖，获得 IAAC 硕士课程奖学金。IAAC MASTER IN ADVANCED ECOLOGICAL BUILDINGS (2019-2020)

备注：对于单元模块奖部分的获奖方案，如涉及实际实施，项目方将会和获奖方案团队就实施方面的事务和细节等进行具体沟通与商议。

2. 竞赛时间表
方案征集开始日期：2018 年 6 月 11 日
参赛报名截止日期：2018 年 8 月 15 日
方案征集截止日期：2018 年 9 月 1 日
初评截止日期：2018 年 9 月 25 日
终评截止日期：2018 年 10 月 15 日

3. 评选机制
分为两轮评选：
第一轮初评：采取线上评选，由国内外专家打分评选，选出 20 组入围方案。
第二轮终评：采取线下评选，由国内外专家现场会议评审，最终评选出大奖方案。

1. Award System
1) First Prize (1 place): 25,000 Euro(Tax included).
2) Second Prize (2 places): 12,000 Euro(Tax included).
3) Unit Module Prize (3 places): 6,000 Euro(Tax included).
4) Finalist Award(20 places): Participants entered in finalist will be awarded Finalist Award Certification.
5) The Next Architecture Scholarship Award: The Jury will select one person who will be awarded a scholarship for IAAC MASTER IN ADVANCED ECOLOGICAL BUILDINGS (2019-2020).

Remarks: For the winning design solution of the Unit Module Prize, the project party will communicate with the awarded team about any issues and details involving actual implementation.

2. Competition Schedule
Start Date of Programme Collections: June 11th, 2018
Deadline of Registration: August 15th, 2018
Deadline for Programme Collections: September 1st, 2018
Deadline for Initial Evaluation: September 25th, 2018
Deadline for Final Evaluation: October 15th, 2018

3. Selection System
Two Rounds of Selection:
- The first rounds of initial evaluation: Experts from The Next Architecture professional committee will score online and select 20 groups of short-listed projects.
- The second round of final evaluation: The final award scheme is selected and will be evaluated on-site by experts.

孟建民
Meng Jianmin

沈迪
Shen Di

李兴钢
Li Xinggang

徐卫国
Xu Weiguo

张雷
Zhang Lei

冯正功
Feng Zhenggong

张彤
Zhang Tong

张应鹏
Zhang Yingpeng

Vicente Guallart

Hernan Diaz Alonso

Daniel V. Hayden

Henriette H. Bier

Michael U. Hensel

大赛评委会
Members of the Jury Committee

大赛邀请以下13位全球嘉宾组成"智慧树——垂直社区的未来生活"国际设计竞赛评审委员会：孟建民：博士；教授；中国工程院院士，全国工程勘察设计大师，深圳市建筑设计研究总院有限公司总建筑师；Vicente Guallart：巴塞罗那前城市总建筑师，巴塞罗那城市生活环境（UrbanHabitat）部门的首位负责人，加泰罗尼亚高等建筑学院（IAAC）创始人；沈迪：全国工程勘察设计大师，教授级高级工程师，华建集团副总裁兼总建筑师，中国建筑学会建筑师分会副理事长；李兴钢：全国工程勘察设计大师，中国建筑设计研究院有限公司总建筑师、李兴钢建筑工作室主持人；Daniel V. Hayden：丹麦 DISSING+WEITLING 建筑事务所合伙人，建筑总监；Hernan Diaz Alonso：南加州建筑学院（SCI-Arc）总监，洛杉矶建筑事务所 Xefirotarch 负责人；张雷：江苏省设计大师，南京大学建筑与城市规划学院教授，可持续乡土建筑研究中心主任，张雷联合建筑事务所创始人兼总建筑师；Dr.-Ing. Henriette Bier：荷兰代尔夫特理工大学建筑与建筑环境学院副教授；代尔夫特机器人研究所，代尔夫特理工大学 Robotic Building Group 的发起人和领导人；RoboticBuilding Lab 主任；张彤：江苏省设计大师，东南大学建筑学院副院长、教授、博士生导师，国际建协教育委员会委员，中国绿色建筑与节能专业委员会绿色建筑设计理论与实践分会副主任委员；张应鹏：江苏省设计大师，九城都市建筑设计有限公司总建筑师，东南大学、浙江大学、华中科技大学兼职教授；Michael U. Hensel：维也纳科技大学数字建筑与规划系主任；OCEAN 设计研究协会主席；前奥斯陆建筑与设计学院 RCAT 实验室主任；徐卫国：清华大学建筑学院建筑系主任、教授、博士生导师，中国建筑学会数字建造委员会副主任，数字建筑设计专业委员会（DADA）专委会主任；冯正功：江苏省设计大师，中衡设计集团股份有限公司董事长、总建筑师。

The Competition invited 13 international experts to form the International Design Competition Jury Committee. Meng Jianmin, Ph. D. Professor, Academician of Chinese Academy Engineering, National Master of Engineering Survey and Design. He is also the Chief Architect of Shenzhen General Institute of Architectural Design and Research Co. Ltd. Vicente Guallart, former chief architect of Barcelona, the first head of the Urban Habitat Department of Barcelona and the founder of the Institute of Advanced Architecture of Catalonia (IAAC). Shen Di, National Master of Engineering Survey and Design, Professor-ranked Senior Engineer. He is the Vice President and Chief Architect of Huajian Group and the Vice President of the China Architectural Society of Architects. Li Xinggang, National Master of Engineering Survey and Design, the Chief Architect of China Architectural Design & Research Institute, the director of Li Xinggang Architectural Studio. Daniel V. Hayden, partner and director of Architecture Affairs of DISSING+WEITLING Architectural Firm in Denmark. Hernan Diaz Alonso, Director of the Southern California Institute of Architecture (SCI-Arc), Head of the Los Angeles-based architecture office Xefirotarch. Zhang Lei, Design Master of Jiangsu Province,. Professor of School of Architecture and Urban Planning, Nanjing University, and Director of Sustainable Local Architecture Research Center. Sustainable Local Architecture Research Center. Zhang Lei is also the founder and chief architect of AZL Architects.Dr.-Ing.Henriette Bier, Associate Professor at the School of Architecture and the Architectural Environment,Delft University of Technology, Netherlands; initiator and leader of the Delft Robotics Institute and the Robotic Building group at Delft University of Technology; Director of Robotic Building Lab. Zhang Tong, Zhang Tong, Design Master of Jiangsu province Professor and Deputy Dean of School of Architecture, Southeast University, Member of UIA Education Commission, Vice Chair, Architectual programming Committee, ASC. Zhang Yingpeng, Design Master of Jiangsu Province, Chief Architect of 9-Town Design Studio for Urban Architecture, Adjunct Professor of Southeast University, Zhejiang University and Huazhong University of Science and Technology. Michael U. Hensel, Director of Department of Digital Architecture and Planning, Vienna University of Science and Technology; Chairman of OCEAN Design Research Association; Former Director of RCAT, Oslo School of Architecture and Design. Xu Weiguo, Chairman, Professor and Doctoral tutor of Department of Architecture, School of Architecture in Tsinghua University, Vice-Chairman of the Digital Construction Committee of the Architectural Society of China, Digital Architectural Design Committee (DADADA). Feng Zhenggong, Design Master of Jiangsu Province, Chairman of the Board and Chief Architect of Zhongheng Design Group Co., Ltd.

什么是下一代建筑？这是一次什么样的竞赛？此次竞赛进行了持续、深入、多种形式的探讨。

What is the "Next Architecture"？What kind of competition is 'Smart Tree'？The competition was conducted in a continuous, in-depth and multi-form discussion.

建筑系统优化篇
Chapter of Architectural Systematic Optimization

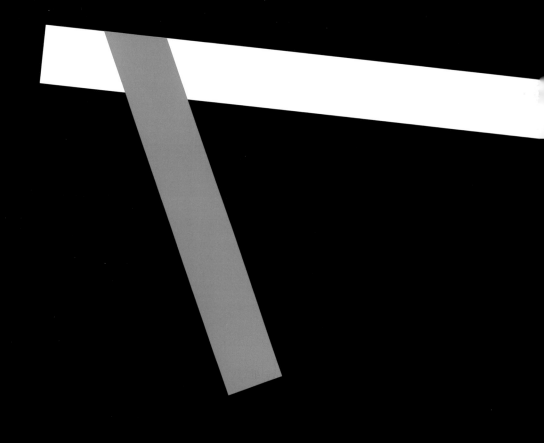

荣获奖项：一等奖　　　　　　Awarded: First Prize

建筑的自然代谢　　BUILDING METABOLISM

团队成员：章阅、郝金立、闫泽明、苗鹤鹏　　Team members: Yue Zhang, Jinli Hao, Zeming Yan, Hepeng Miao
来　　自：沈阳建筑大学　　　　　　　　　　From: Shenyang Jianzhu University

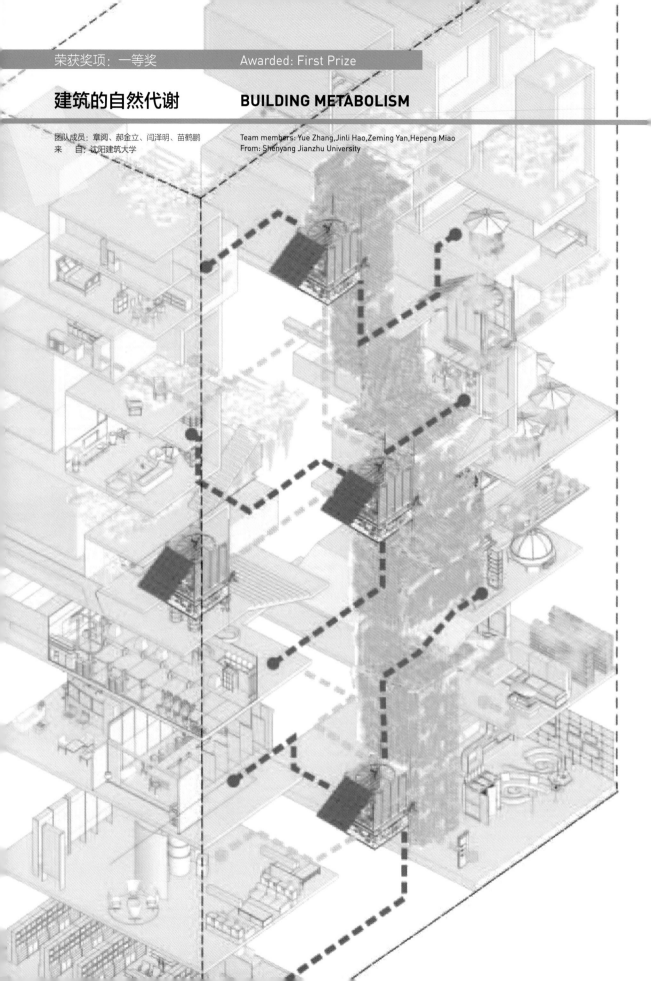

概念
CONCEPT

在本次设计中，我们将建筑物视为一个有机体，类比生物学的胞吞胞吐的现象，置入自然代谢平衡单元模块。该模块作为建筑生态系统的一部分在时刻运动着，在需要搜集废弃物时，资源代谢装置通过"胞吞"将太阳能转化为电能供自身使用，并采集人们的厨余垃圾、代谢废弃物和可回收垃圾放在微生物分解箱内，进行一段时间的降解处理后，进行胞吐，释放出人们生活所需要的可燃性气体——CH_4，以及对环境友好、植物需要的水和二氧化碳，通过这个可持续的动态循环过程，形成一种新的建筑内分布的动态能量循环体系。同时，该模块也是一个移动式菜园和休憩空间，可通过APP与人们产生在线互动与上层的居住单元模块拼接产生新的活动空间。

In this design, we regard buildings as an organism, make analogy between them and the biological phenomenon of endocytosis and exocytosis, and set the natural metabolic balance unit module. As part of the building ecosystem, the module is on the move all the time. When waste needs to be collected, the resource metabolism device converts solar energy into electric energy for its own use through "endocytosis" and collects people's kitchen waste, metabolic waste and recyclable waste in a microbial decomposing box. After a period of degradation treatment, the module releases combustible gas, methane (CH_4), water and carbon dioxide which are environmentally friendly and are needed by the plants. Through this sustainable dynamic circulation process, a new distributed dynamic energy cycle process is formed in the building. At the same time, the module is also a mobile vegetable garden and recreational space, which can encourage interaction between people through an online APP while making new living spaces by splicing with the upper living unit module.

一般能量流动
GENERAL ENERGY FLOW

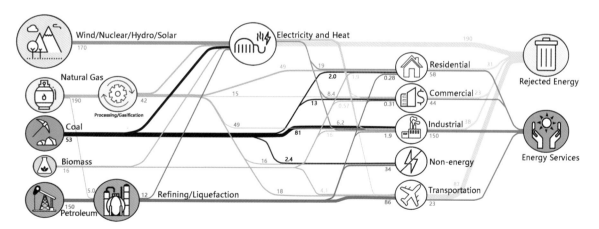

此设计的背景是人类参与下的自然界的一般能量流动，尽管能量在整个宇宙中是守恒的，物理学有能量守恒定律，但是人类在使用能量的时候，从采集自然资源，再到为己所用，过程中会有很大的损失，而且是一个单向不可逆的过程。我们现代的大部分建筑也存在同样的问题，需要消耗大量的能源来维持建筑内部系统的运行。

This design was born with the concept of the general flow of energy in nature with human participation. Although energy is conserved in the whole universe according to the law of Conservation of Energy, when we use energy, there will be a great loss in the process from collecting natural resources to using them, which is a one-way irreversible process. Most of our modern buildings have the same problem and need to consume a large amount of energy to maintain the operation of the internal system of the building.

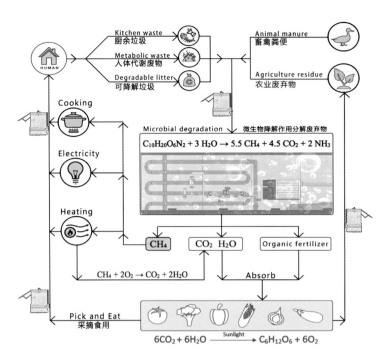

通过引入自然代谢平衡单元模块，使建筑物在内部进行能量循环，在随着季节变化更替之时平衡建筑内在生态系统成为了可能，使建筑与自然有了交流，形成内在生态平衡和自我调节机制，使建筑变得智慧化和生命化。

By introducing the natural metabolic balance unit module, it makes it possible for the building to carry out energy circulation inside, while balancing the building's internal ecological system with the change of the seasons; thus making the building communicate with nature, forming an internal ecological balance and self-regulation mechanism while resulting in a comprehensive intelligent and vital living place.

自然代谢平衡单元模块剖面图
SECTION VIEW OF NATURAL METABOLIC BALANCE UNIT MODULE

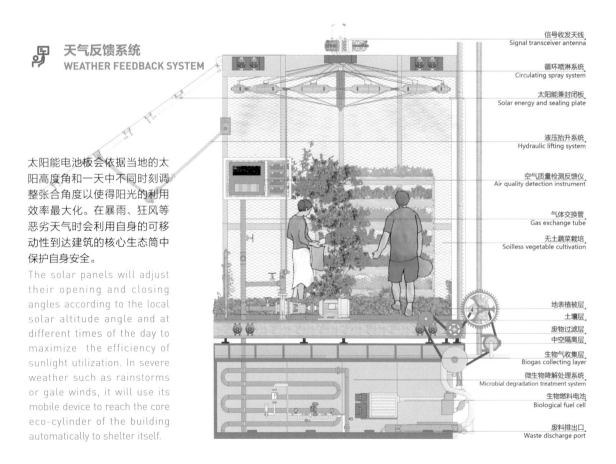

天气反馈系统
WEATHER FEEDBACK SYSTEM

太阳能电池板会依据当地的太阳高度角和一天中不同时刻调整张合角度以使得阳光的利用效率最大化。在暴雨、狂风等恶劣天气时会利用自身的可移动性到达建筑的核心生态筒中保护自身安全。

The solar panels will adjust their opening and closing angles according to the local solar altitude angle and at different times of the day to maximize the efficiency of sunlight utilization. In severe weather such as rainstorms or gale winds, it will use its mobile device to reach the core eco-cylinder of the building automatically to shelter itself.

响应人的需求
RESPOND TO PEOPLE'S NEEDS

单元模块不仅可以提供人们生活所需的绿色蔬菜，还可通过 APP 与居民在线互动连接，例如 90 后会出现长期宅在室里的情况，需要内部空间满足健康的要求，也就是其环境的调节具有可变性，通过人工智能的观察，出门天数较少的人们系统会自动派遣该代谢模块与其居住模块相连以保证人的健康。

The unit module can not only provide green vegetables for people, but also interact with residents online through APP. For example, Post-90s generation tends to stay in the room for a long time and needs internal space to meet the need of exercise, which requires the adjustment of their environment. Through the observation of artificial intelligence, people who don't have enough days to go out will automatically be sent the metabolic module to connect with their living module to ensure their health.

 能源生产储存
ENERGY PRODUCTION AND STORAGE

利用自身的生物燃料电池和生物气产出清洁无污染的能源，在移动到各家各户时供给人们使用，让能量在一栋建筑内部循环成为可能。

Clean and pollution-free energy produced with its own bio-fuel cell and bio-gas can be supplied to every house for people to use, making it possible for energy to circulate inside a building.

 废物收集运输
WASTE COLLECTION AND TRANSPORTATION

城市垃圾被运往其他地方的垃圾填埋场。这种方法只是转移了问题没有认识到浪费过多的原因不仅仅是城市的庞大规模，而且主要是消费者一次性文化的结果。该模块为尽可能多的生活用品创建闭环循环，借此唤醒人们的环保意识。

Traditionally, city garbage is transported to landfills elsewhere, in which this method only shifts the problem and contributes to excessive waste. People don't realize it partly because of the huge scale of the city and partly because of the result of the disposable consumption culture found in consumers. This module creates a closed loop for as many household goods as possible, which is designed to awaken people's awareness of environment protection.

 联系整个系统
CONTACT THE WHOLE SYSTEM

作为一个建筑物内的活动体，该模块承担了运输者、分解者、生产者的职能，同时也是建筑的传感系统，亦作为一个机械系统的执行者存在着，同时也是居民生活活动的重要载体，在整个系统中起到了不可或缺的作用。

As a moving body in a building, the module takes on the functions of transporter, decomposer and producer. As it is the sensing system of the building as well as the executor of a mechanical system and an important carrier of residents' life activities, it plays an indispensable role in the whole system.

自然代谢平衡单元模块系统拆解图
DISASSEMBLING DIAGRAM OF THE NATURAL METABOLIC BALANCE UNIT MODULE SYSTEM

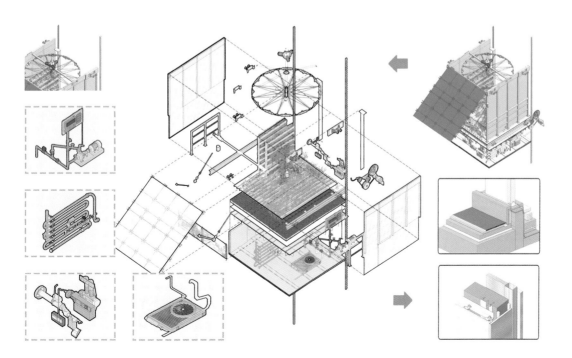

自然代谢平衡单元模块与住宅单元模块的信息联系
INFORMATION RELATIONSHIP BETWEEN NATURAL METABOLIC BALANCE UNIT MODULE AND RESIDENTIAL UNIT MODULE

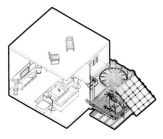
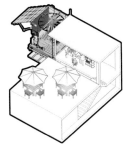

住宅单元由3m×3m×Nm基本模数单元组合而成，自然代谢平衡单元模块的尺寸为3m×4m×4m，可与住宅单元拼接。居民可通过APP与代谢平衡模块相连互动，此时居民的手机为传感系统，发送的信息由建筑的中枢控制系统进行运算判断，再传递信息到机械系统，代谢平衡模块便会移动到与居民住宅单元模块相对应的入口处供其使用。

The residential unit is composed of 3m×3m×Nm basic modular units. The size of the natural metabolic balance unit module is 3m×4m×4m, which can be joined with the residential unit. Residents can interact with the metabolic balance module through an APP, in which case the residents' mobile phones become a sensing system, and the information sent out is calculated and judged by the central control system of the building, and then transmitted to the mechanical system. Eventually, the metabolic balance module will be moved to the entrance corresponding to the residential unit module for use.

生态模块与住宅单元系统组织关系
THE ORGANIZATIONAL RELATIONSHIP BETWEEN ECOLOGICAL MODULE AND RESIDENTIAL UNIT SYSTEM

下一代建筑不仅仅是单独的个体，而是整个环境集合里的元素，和大型生态网络里的一个单元体。以单元为模块在三维空间尺度上进行移动、交换、打破传统固定单一垂直模式，以智慧学习控制为原点辅以X、Y、Z坐标体系，各个方向发散，以探求建筑空间的可变性。

The 'Next Architecture' does not exist alone, it is also one of the elements in the whole set of environments and one of the units in a large ecological network. In order to explore the variability of architectural space, every unit used as a module can be moved and exchanged on the scale of three-dimensional space and can diverge in all directions by using intelligently learning controls as the origin with X, Y and Z coordinate systems, breaking the traditional fixed single vertical mode.

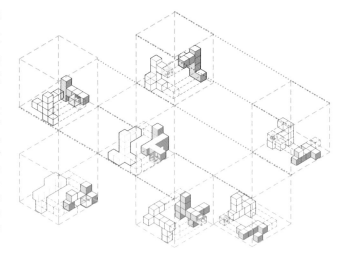

垂直交互空间探讨
DISCUSSION ON VERTICAL EXCHANGE SPACE

居住者可根据自己的需要，通过手机APP选择不同模块，订制属于自己的居住空间，亦可与自然代谢平衡单元模块产生互动，此时的建筑便不再是居住的容器，而成为人机互动桥梁，建筑由一开始的生态能量循环基础设施构架逐渐生长变化，成为多功能多元的集合体。每个居住者或者群体可以按照主观意志选择模块，然后重新自定义其功能与空间形态，从而满足个体差异化的空间需求。

Residents can choose different modules according to their own needs through the mobile APP to customize their own living space or interact with the natural metabolic balance unit modules. In this way, the building is no longer a container for living, but a bridge for human-computer interaction. Since the beginning, the building has gradually grown and developed into a multi-functional, comprehensive residential quarter from the initial ecological energy cycle infrastructure. Each resident or group can choose a module freely according to their own will and then redefine its function and space forms to meet their individually differentiated space needs.

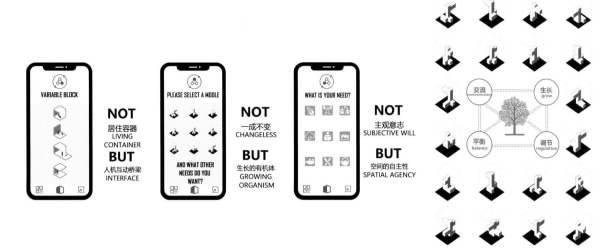

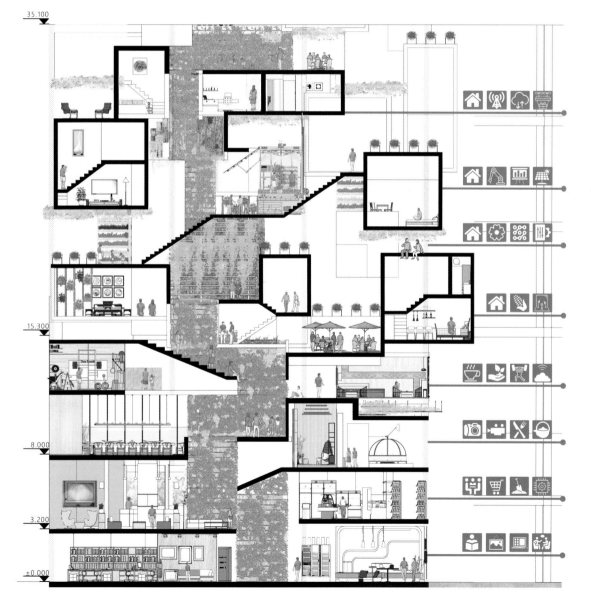

室内——室外（住宅单元）空间类型
INDOOR-OUTDOOR (RESIDENTIAL UNIT) SPACE TYPE

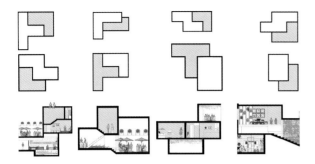

住宅单元由3m×3m×Nm基本模数单元组合而成，方块之间的组合与离散形成了丰富的室内与室外空间，近距离享受共享空间。其中每户设有过渡性的半私密空间与之衔接。

Residential units are composed of 3m×3m×Nm basic modular units, and the combination and dispersion of newly-formed units provides an abundance of indoor and outdoor space, enabling people to enjoy the shared space while being close to each another. Each household has a transitional semi-private space to connect with each other.

居住单元模块组合形式
COMBINED FORM OF RESIDENTIAL UNITS

针对入住人群具有流动性、租赁性的特征，套型以标准化、模块化、可变性为原则，套型可合可分。

In view of the characteristics of mobility and the occupants' ability to rent, the flats are standardized, modularized and adaptable so that the ones on the same floor can be joined and separated accordingly.

独处　　观赏　　通过　　共享

单人居住　　双人居住　　双人居住　　三人居住

技术图纸
TECHNICAL DRAWINGS

16.2 m 平面图 1:100
16.2 m PLAN 1:100

25.3 m 平面图 1:100
25.3 m PLAN 1:100

21.8 m 平面图 1:100
21.8 m PLAN 1:100

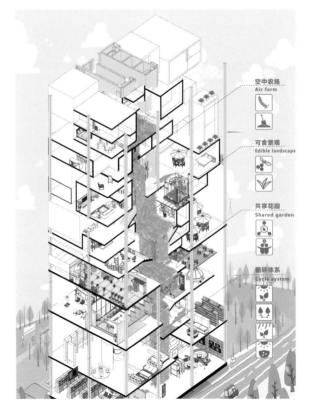

18.3 m 平面图 1:100
18.3 m PLAN 1:100

28.4 m 平面图 1:100
28.4 m PLAN 1:100

31.4 m 平面图 1:100
31.4 m PLAN 1:100

结语
CONCLUSION

我们试图将下一代建筑作为有机体来理解，用科技的方案实现建筑内能量与物质的动态循环，通过自下而上的群体智慧实现未来建筑空间的自组织，正如同电影《黑客帝国》中影片末尾救世主对人工智能的控制者说道 "一个充满可能性的世界。以后的发展就让他们自己决定。"在这样的世界里，每个人都有能力去创造属于自己的自由和全知全能。

We attempt to gain an understanding of the 'Next Architecture' as an organism, in order to realize the dynamic circulation of energy and matter in the architecture, on the basis of which, through bottom-up group intelligence, the self-organization of future architectural space will come true with the aid of a scientific and technological scheme. Just as the one——Neo said, towards the end of the movie The Matrix, to the controller of AI, "A world where anything is possible. Where we go from there is a choice I leave to you." In such world is the future where everyone will be able to freely create their own freedom and omnipotence.

荣获奖项：入围奖　　　　　　　Awarded: Finalist Award

漂浮的原野
——生长的未来居住空间

SUSPENDED FIELDS —— Future Living Space for Growth

团队成员：刘松，孙瑜，刘晓梅，杨卉，邬志恺，季霏
来自：中设设计集团股份有限公司

Team members: Liu Song, Sun Yu, Liu Xiaomei, Yang Hui, Wu Zhikai, Ji Fei
From: China Design Group Co., Ltd

在人类文明的历史长河中，住房已经经历了三代，第一代茅草房，第二代砖瓦房，第三代电梯房。他们通过窗户看到外面世界，呼吸新鲜空气，推开门和邻居亲切互动，和大自然亲密接触。显然这种居住状态不是人类的理想模式，人类的理想住宅是什么样的？

人类起源于水，鱼栖于水，人居于陆，通过观察鱼类的生存环境，我们设计并制作生态鱼缸，铺泥、施肥；石、木造景；种植水草；照明系统；过滤系统；智能化控制系统；注水开缸。

那么什么是人类更宜居的2.0生活模式？

首先，居住应该与自然融合，环境与奢华同在，人在庭院中穿行，每层楼就是一个合院，每户都带有一个私家花园，建筑外墙有绿植围护，人与自然和谐共生。

In the long history of human civilization, architect has gone through three generations, the first generation is thatched houses, the second one is brick houses, and the latest generation is apartment with elevators. They open their windows to see the outside world and to let the fresh air in. Also, they open the doors and interact intimately with their neighbors, and keep close with nature. Obviously, this living condition is not the ideal mode of human living. What is the dream residence of human beings?

Human beings originate from water. As we all know, fish live in water, people live on land. By observing the living environment of fish, we design and make ecological aquarium by a series of work: paving mud and fertilizing, stone and wood landscaping, planting aquatic plants, building lighting and filtering system, controlling the system intelligently and injecting the water and opening the aquarium.

So what is the Generation 2.0 life style of human habitat?

First of all, living should coexist with nature, environment and luxury. Each floor is equipped with a courtyard for people to walk in. Each household has a private garden. Green fences surround the building. In general, people and nature live in harmony.

中国，南京，江北新区国际健康城人才公寓3号楼。设计范围：建筑地面层以上 35.1 米高的区段。开放式的庭院设计，赋予空间组织的灵活性与多变性，多层错落的组合设计，表达对生活的创意拼装和对家的无限遐想。
Building No. 3 of the International Healthcare Area Talent Apartment in Jiangbei New Area, Nanjing, China. Scope of design: 35.1 meters above the ground floor. Open-style courtyard design provides space organization with flexibility and variability, while multi-layered and scattered combination design express the creative assembly of life and the infinite reverie of home.

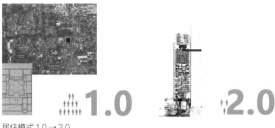

居住模式 1.0 → 2.0
Residence pattern 1.0 → 2.0

设计通过条状的建筑形式来体现 Box Hill 的多元文化背景。建筑的"条带状"结构几何交错地覆盖在楼板上，形成一种连贯的立面形态。"条带状"的末端采用彩色的饰面装饰阳台，以此来增加公共区域的活力。同时，为了增加公寓居住者的舒适感，内部流通空间向外部开放，作为一个建筑的"肺"来打造。这个"肺"将提供更多的采光和自然通风，同时运用植物来装饰开放的庭院。
The design reflects the multicultural background of Box Hill through a strip form of architecture. The "strip" structure of the building is geometrically interlaced to cover the floor, forming a coherent vertical form. The "strip" ends are decorated with colored faceplates to enhance the vitality of public areas. At the same time, in order to increase the comfort of apartment dwellers, the internal circulation space is open to the outside, as to build a "lung" part. The lung will provide more daylight and natural ventilation while using plants to decorate open courtyards.

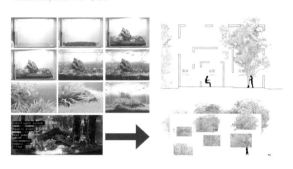

在公共单元中，围绕中庭空间，打造架空绿色书吧、生态咖啡厅、天际架空健身房、林间画室、冷链超市、创客空间等，构建绿色社交的生活场景，让业主与大自然和谐共处，传递着全新的生活方式。
In the public unit, around the atrium space, the green book bar, ecological cafe, skyline overhead gymnasium, forest studio, cold chain supermarket, makerspace, etc., together create a green social life scene, so that owners and nature could live in harmony, passing on a new way of life.

绿植顺着建筑外墙垂直生长，"漂浮的原野"应运而生。从建设之日起就彻底拒绝城市病的发生。垂直绿墙、空中花园和屋顶花园系统，形成天然屏障，起居空间、办公空间与景观相互交替，伴着流动的空气、温暖的阳光、吹拂的微风，人的情感在环境变换中得到滋润。
Green plants grow vertically along the external walls of the building and "suspended fields" emerge as the times require. From the beginning of construction, we completely reject the occurrence of urban diseases. Vertical green walls, aerial gardens and rooftop gardens form a natural barrier. Living space, office space and landscape alternate with each other. With the flowing air, warm sunshine and blowing breeze, people's emotions are nourished in the changing environment.

如何让人类生活模式从传统建造向智能化、定制化更迈进一步？将可变的建筑硬件与智能的控制软件相结合，在有限的实体空间里，拓展出无限的居住可能。通过各类智能化系统的引入，让人类的居住、办公需求模式存于芯片或云端，随时载入，让空间整体充满艺术性与想象力，体验智能化生活时代。

How can the human life mode move from traditional construction to an intelligentized and customized one? The combination of variable building hardware and intelligent control software expands the infinite living possibilities in the limited physical space. Through the introduction of various intelligent systems, the living and office modes of human beings can be stored in the chip or the cloud, which can be loaded at any time, so that the overall space is full of artistry and imagination, and people can experience the era of intelligent life.

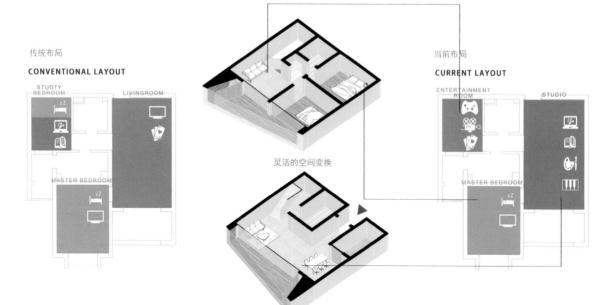

人类社会越进步，就会越贴近大自然，漂浮的原野让你置身自然之中、配以智能化的生活的模式，这也许就是下一代建筑，垂直社区的未来生活方式！

The more progress the human society makes, the closer it will be to the nature. The suspended fields enable you to be as in the nature, accompanying with the intelligent life mode. And this may be the future life style of the next generation of buildings and vertical communities!

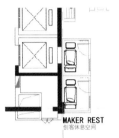

MAKER REST
创客休息空间

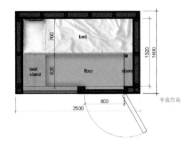

平面方案

效果图

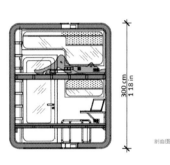

剖面图

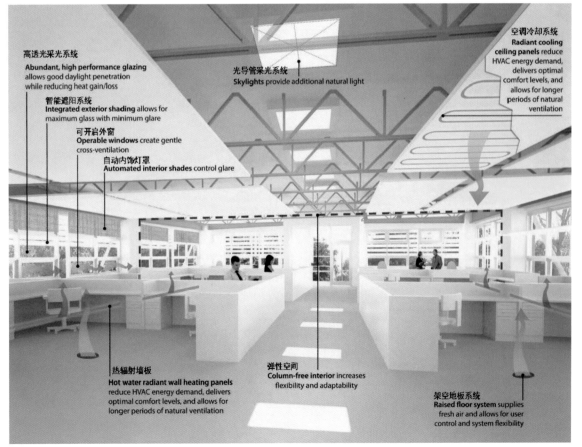

孟建民
Meng Jianmin

博士；教授；中国工程院院士
深圳市建筑设计研究总院有限公司总建筑师

Ph. D, prof., and academician at the Chinese Academy of Engineering, Principal of Meng Architects of Architectural Society of China

科技拓展建筑的边界
Expand the Boundaries of Architecture via Science and Technology

此次"智慧树：垂直社区的未来生活"建筑竞赛项目，开启了对建筑未来性的探索。竞赛中的部分作品对于科技与建筑结合等方面具有一定的启发性，为我们探索下一代建筑提供了一些新鲜的思路和手段。随着今后类似活动的开展，建筑领域、科技领域乃至整个社会，都将进一步关注到未来建筑及未来城市规划的发展走向。科技发展日新月异，不仅重新定义了人的生活方式，也重新定义了我们对城市与建筑的需求。同时，科技创新也为城市建筑的发展提供了相应的技术支持。

The competition "Smart Tree: Future Life in a Vertical Community" marked the start of our exploration on possibilities of futuristic architecture. Some participants' solutions shed light on the integration of technologies and architecture among others and provided us with new ideas and methods regarding the next architecture. With more similar projects and events to come, people in the architecture industry, scientific and technological fields and even society would pay more attention to the developmental direction of architecture and the city. Fast developing science and technologies have redefined people's way of living as well as our needs for cities and buildings. Meanwhile, scientific and technological innovations also provide technological support for the development of buildings in cities.

在城市传统高密度核心区，不可避免存在人口密集、土地稀缺、交通拥堵、能耗较高及环境不佳等问题。城市和建筑发展，必须以人为本，找到城市核心痛点，再解决发展路径问题。在建筑设计和城市规划中，不仅需要遵循传统的设计思路，更要综合考虑设计本身与其他资源的整合、调配及可持续发展。

In urban high-dense core areas, there are inevitable existing problems including high population densities, land scarcity, traffic jams, high energy consumption and unfavorable living environments. The development of cities and architecture should be based on people's needs. We should first solve the core problems in urban development and then find the right path of development. In architectural design and urban development, we should not only follow traditional ways of design, but also factor in consolidation, coordination and sustainable development of design and other resources.

未来的城市，不应就设计谈设计，就规划谈规划。同样，建筑学的边界将在科技创新的影响下越来越模糊，不再受限于风格、类型等传统的概念。下一代建筑，将在创新技术支持之下实现智慧化、动态化的发展，其多样的功能与形态将融入人类的各种生存形式中：建筑将不再是静态的立面，而能以扭转移动的形体为人类带来动态的体验；也不再局限于僵化的外观，而是充满智慧的伙伴，成为一艘能全方位协助生活应用的诺亚方舟。

Design and planning of cities in the future should not be restricted. Boundaries of architecture would become increasingly vague with the influence of scientific and technological innovation. And architecture would not be restricted by traditional concepts like styles and categories. Architecture of the next generation would become more intelligent and dynamic with the support of innovative technologies. And diversified functions and forms of architecture would be integrated into all aspects of people's lives; architecture would no longer be static, as it would be a twistable, movable and dynamic environment in which it can provide for people. Moreover, buildings would no longer be in a fixed form, but would become intelligent partners that can facilitate people's lives in all aspects.

在遥远的将来，我们的城市将不再依赖简单的规模扩张，而是伴随着科技与建筑边界的拓宽，在新的时代实现建筑与人类的交融合一。

In the distant future, urban development would no longer rely on scale expansion. As boundaries of technology and architecture expand; architecture and human lives will be integrated together in the new era.

张彤
Zhang Tong

东南大学建筑学院副院长、教授、博士生导师
国际建协教育委员会委员
中国绿色建筑与节能专业委员会绿色建筑设计理论与实践分会副主任委员
江苏省设计大师
Professor, Deputy Dean, Doctoral Tutor of Architecture School in Southeast University
Member of UIA Education Commission
Vice Chair, Architectural Programming Committee, ASC
Design Master of Jiangsu Province

未来建筑的智慧变革
Intelligent Revolution of Buildings of the Future

房屋建筑构成了城市和社会基底。人类的行为与其构成的历史无不发生在建筑承载的环境中，而建筑本身在此过程中却表现得相对稳定，甚至很难察觉它的变化。而一旦建筑开始产生变化，那一定代表着整个社会已经发生了不可逆转的变革。就目前的现实和未来的趋势看，对下一代建筑产生的重要影响来自两种技术——信息技术和环境技术。20世纪后半叶开始，人类逐渐意识到环境危机和对自身的影响。社会的增长在空间和资源上是有限度的，这一认识改变了以往的思维模式以及我们生产和生活。与此同时，包括计算机、互联网、物联网、大数据和人工智能在内的信息技术发展迅猛，我们甚至还来不及适应，技术本身已经迭代革新，进而改变社会的组织和生产模式。基于这样的认识，我认为对未来建筑最重要的影响将来自于环境技术和信息技术，作为社会基质和承载的建筑，也会因这两类技术而发生变革。

Buildings are the infrastructure of our city and society and also act as the environment for all human behaviors and events in history. Building itself is relatively stable and its changes are hardly noticeable. Yet when they do change, it surely indicates the society have undergone irreversible transformation. Current and future trends suggest that information technology and environmental technology would have great influence on architecture of the next generation. Since later half of the 20th century, humans have gradually recognized the existence of environmental crisis and its influence on their life. Social growth is limited in terms of space and resources, and this recognition has changed the previous mode of thinking and our production and life.Meanwhile, computers, Internet, IoT, big data and AI among other information technologies are developing in such a fast speed that we can barely adapt to them before new iterations and innovations are in place. Consequently, society organizational structure and production models have also changed. Considering this, I think information technology and environmental technology would have the greatest impact on architecture in the future. Buildings, the fundamental environment of our society, are sure to change due to these two kinds of technologies.

建筑学的核心价值曾经在于其清晰、稳定、恒久的结构，而这一点在当今正在发生变化。构筑传统房屋的是砖瓦——一种物质材料。进入信息社会后，构筑房屋的将是数据。数据已成为这个社会最核心的资源。也正因此，未来建筑会在三个方面产生变化。

The core value of architecture lies in its clear, stable and everlasting structure. Yet this is changing now. Traditionally, houses are built by bricks, which are tangible materials. In the era of information, buildings would be made of data. Data would become the most important information in the world. Thus buildings of the future would change in terms of three aspects.

首先，从固定结构向柔性结构的转变。相对于传统建筑固定静止与隔离的特性，下一代建筑将呈现出开放、可变和交互的柔性。造房子的原始动机，是从外部世界隔离出一个自我的内部，构筑一个身体的庇护所，进而寻求抵抗时间的永恒性。当代信息技术的飞速发展，使我们的生活充满了变化的可能。当无需更换地点，便可在同一时间切换不同场景、调用各种资源时，房屋结构的可变性和交互性将成为核心特征。墙体不再静态，而会承载信息、记录数据，并随时切换情景模式和信息呈现，通过交互满足生活需要。房屋不再是一个把我们从外部世界隔离开来的庇护体，我们身处于由各种信息服务网络构成的社会中，我们的房屋将成为信息、服务与资源网络的终端，正如同手机是网络信息的智能终端一样，建筑会成为生活服务的智能化终端。

First, change from fixed structures to flexible ones. Traditional buildings are relatively quiet and exclusive, while those of the next generation would be open, flexible and interactive. In the beginning, we build houses to detach ourselves from the external world and regard them as our shelter. We hope they could protect us for a very long time. However, as information technologies develop at full speed, our life keeps changing. When it is viable to shift scenarios and allocate various resources without changing positions, flexibility and interactivity of building structure would play a vital role. Walls would no longer be static. They would convey messages, record data, support profile switching and information displaying at any time. They need to interact with people to meet their needs. Houses are no longer shelters that detach us from the external world. In a society formed by various information service networks, houses would become terminals for the network of information, services and resources. As mobile phone is the intelligent terminal of network information, building would become an intelligent terminal of services in people's everyday life.

第二个变革来自人工智能。建筑将能否成为具有自我学习能力的智慧体？我认为完全可能。如果以深度自学习来定义人工智能，未来建筑亦可掌握这种获取与分析信息、制定与执行策略的能力。我们可能会面对一个有感情、有思想、有行动力的房屋，而不是一个沉默的对象，它会主动基于数据处理来改变空间与生活模式，从而更高效地与人互动。它甚至可能不再是一个无机世界的对象，而是一个人工智能体。不管我们是否做好准备，都可能会面对这种挑战——与具有生命体特征的建筑共同生活。

The second reform is enabled by artificial intelligence. Will buildings become intelligent beings capable of learning by themselves in the future? I think this is highly possible. As artificial intelligence features deep learning, buildings may be able to use this method to acquire and analyze information, draft and execute strategies. Houses may have their own emotions, ideas and may take actions by themselves. They would no longer remain silent. Houses may change space structures and living modes based on data analysis results, thus interacting with people more efficiently. House could even become an artificial intelligent being instead of an inorganic one. Whether we are prepared or not, we may face the challenge to live in buildings with features of a living creature.

最后，建筑将成为回归自然的智慧生态系统。回到环境问题，一直以来，我们都把自己与自然设立为二元对立体。这种躲避、改变和控制自然的态度将会改变，我们更希望与环境成为一体。如果个体的房屋具有了生命体机制，整个社区和城市是否可以想象为一个更大的智慧系统？随着环境技术与信息技术的发展，海量数据的采集、汇集和云端操作，实现更大量数据的复杂运算将使人工建造系统与原生自然界的一体化成为可能，也为构筑"城市山林"这种更大更复杂的智慧体提供了可能。最重要的是，这个由建筑组成的人工生命机制与自然界的生命机制相吻合，而非二元对立，从而创造出人与自然共融共生的人因智慧生态系统。

Last, buildings would become smart ecosystems that allow people to return to nature. As for environmental issues, we always look at human and the nature from the perspective of binary opposition. In the future, we would no longer want to evade, change and control the nature like today, and we would like to integrate with the environment. If the house has the mechanism of a living creature, could the entire community and city be regarded as a large intelligent system? With the development of environmental technology and information technology, artificial construction system could integrate with the nature when massive data could be collected, gathered and processed via cloud services, and more complex computation of even larger quantities of data can be conducted. This also makes it possible to build a larger and more sophisticated intelligent ecosystem of "hills and forests within cities". Most importantly, this artificial life mechanism of buildings is in line with nature instead of a relation of binary opposition. As a result, we can create an intelligent ecosystem in which people and nature coexist in harmony.

建筑生态创新篇

Chapter of Architectural Ecological Innovation

荣获奖项：二等奖　　　　　Awarded: Second Prize

云舟　　　　Cloud Boats

团队成员：董升浩、沙金、和子珉、王浩凡、赵伟、王天鹤、司徒一成、陈睿杰、肖建峰、王言予、王丽娜、李继先、王文胜
来自：云南城投置业股份有限公司、北建院约翰马丁国际建筑设计有限公司 Odesign 工作室、上海云十科技公司、南京华磁科技实业有限公司

Team members: Dong Shenghao, Sha Jin, He Zimin, Wang Haofan, Zhao Wei, Wang Tianhe, Situ Yicheng, Chen Ruijie, Xiao Jianfeng, Wang Yanyu, Wang Lina, Li Jixian, Wang Wensheng
From: Yunnan Chengtou Real Estate Co., Ltd. Product R&D Center, Beijianyuan John Martin International Architectural Design Co., Ltd. Odesign Studio, Shanghai Yunshi Technology Co., Ltd.,Nanjing Hua Magnetic Technology Industrial Co., Ltd.

"我们的成果绝对不是解决某个实际问题的，也不是对于某个技术方向的思考，它应该是更加宏观的，更能引发深思的方案，它能让人们思考到更深的精神层面与社会层面以及人对于自然的追寻本能，让人感知到建筑也可以跨越学科和思维界限。"
"Our results are definitely not to solve a practical problem, nor to think about a certain technical direction. It should be a more macroscopic and more thought-provoking solution. It can allow people to think about the deeper spiritual and social aspects as well as people's human instinct to pursue nature; it is perceptible that architecture can also cross the disciplines and boundaries of thought."

下一代建筑与山水
"Next Architecture" and Landscape

未来北京
The Future in BEIJING

未来上海
The Future in SHANGHAI

未来南京
The Future in NANJING

概念
CONCEPT

如果说古典的城市是关于神话的，现代城市是关于资本和权力的，那么未来的城市就应该是关于人与自然的。
If the classical city is about God and the modern city is about capital and power, then the future city should be about people and nature.

人在朝朝暮暮，山山水水，风风雨雨，一草一木天地之间无不有感而发，触景生情。砖头瓦块，玻璃钢铁，是情、境中的载体，为屋为市，称之为诗意和意境。
People are easily touched by time, natural landscape and phenomena, and every little things on the earth. Bricks, tiles, glass, steel and iron are the carriers of the emotion and environment, form the houses and the cities, and are thus called as poetic flavor and artistic conception.

未来城市将要回到自然的怀抱，是诗意的，是人，人性的城市。
The city of the future will return to the embrace of nature. It is a city of beauty-appreciation, humanity and human nature.

建筑服务器
BUILDING SERVER

装配式 1.0 时代
ASSEMBLY 1.0 ERA
2015 年
THE YEAR 2015

2015 年末发布《工业化建筑评价标准》，决定 2016 年全面推广装配式建筑。大量的建筑部品由车间生产加工完成，构件种类主要有：外墙板、内墙板、叠合板、阳台、空调板、楼梯、预制梁、预制柱等。

The《Industrial Building Evaluation Standards》was released at the end of 2015, and it was decided to fully promote the assembly buildings in 2016. A large number of building parts are produced and processed by the workshop. The main types of components are: exterior wall panels, interior wall panels, laminated panels, balconies, air conditioning panels, stairs, prefabricated beams, prefabricated columns, etc. The assembly type mainly stays on the basic structure and metal components of the building.

集成化 2.0 时代
INTEGRATED 2.0 ERA
2018 年
THE YEAR 2018

已经出现以交通核为核芯的集成模块，核芯主要集成服务空间以及设备管线，室内空间围绕核芯自由灵活布置，降低了空间复合可变成本。

The integrated module based on the traffic core has emerged. The core mainly integrates the service space and the equipment pipeline. The indoor space is freely and flexibly arranged around the core, which reduces the variable cost of the spatial composition.

集成化 + 框架 3.0 时代
INTEGRATED + FRAMEWORK 3.0 ERA
2023 年
THE YEAR 2023

所有建筑房间都可以通过装配式组装实现连接使用，每个功能房间都可以灵活变动以及拆装，可根据需求订购替换功能房间。

All construction of rooms can be connected by assembling. Each functional room can be flexibly altered and disassembled. The replacement function room can be ordered according to requirements.

集成化 + 框架 + 移动 4.0 时代
INTEGRATED + FRAMEWORK + MOBILE 4.0 ERA
2050 年
THE YEAR 2050

居住模块与建筑完全实现装配化生产，居住模块连入自动驾驶技术进行行驶操作，建筑内不是唯一的居住地，居住模块通过可接驳式连接与建筑联通。建筑不再需要围护结构，转变为停靠居住模块的综合服务器。

The residential module and the building are completely assembled and produced. The residential module is connected to the automatic driving technology for driving operation. The building is not the only place of residence. The residential module is connected to the building through a connectable connection. The building no longer needs the envelope structure and is transformed into a comprehensive server that docks the residential modules.

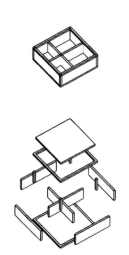
装配式 1.0 时代
Assembly 1.0 Era

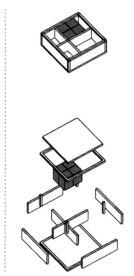
集成化 2.0 时代
Integrated 2.0 Era

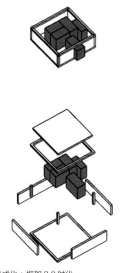
集成化 + 框架 3.0 时代
Integrated + Framework 3.0 Era

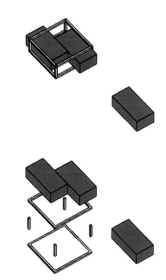
集成化 + 框架 + 移动 4.0 时代
Integrated + framework + mobile 4.0 Era

O pro
O pro

通过生活单元的提取和使用，连入自动驾驶和能源处理系统，成为新一代的交通工具 O pro，形成居住 + 交通新一代居住方式，人们可以按照自己的喜好来选择居住地，不需要固定居所，形成新一代游牧生活，极大减少空间和设备浪费。

Through the extraction and use of living units, it is connected to the automatic driving and energy processing system, becoming a new generation of transportation O pro, forming a new generation of residential + transportation. People can choose their place of residence according to their own preferences, without the need for a fixed residence. Forming a new generation of nomadic life, greatly reducing space and equipment waste.

飘浮运动，无人驾驶
FLOATING, UNMANNED

集工作、居住一体，其场景随意切换，智能家居，实现无家具化。模块化组合，厨卫模块与核心筒服务运转对接。

It integrates work and residence, and its condition is switched freely and the smart home can be actualized furniture-free. A modular combination of the kitchen and bathroom modules works in tandem with the core tube service.

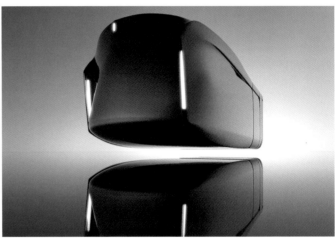

城市连接性
URBAN CONNECTIVITY

无人驾驶汽车可能会在基础设施欠发达的城市和地区产生更大的影响。如纽约、伦敦、巴黎、东京、新加坡、北京、上海等城市是高密度发展的，它们应该更多的鼓励发展公共交通，以保持现有的道路空间。城市内通达性较差的区域，例如发达城市的郊区和欠发达地区的城市，将会受到更多来自无人驾驶汽车和城市交通设施一体化的影响。

另一方面，不论城市区域的连接性强弱，联网的无人驾驶汽车都将成为这个城市的重要组成部分。发达城市的核心区将会以无人共享汽车和无人驾驶巴士的形式运行。

Driverless cars may have a greater impact in cities and regions with less developed infrastructure. Cities such as New York, London, Paris, Tokyo, Singapore, Beijing, and Shanghai are developing at a high density. The cities should encourage more development of public transportation to maintain the existing road space. Areas with poor accessibility within urban areas, such as the suburbs of developed cities and cities in underdeveloped areas, will be more affected by the integration of driverless cars and urban transportation facilities.

On the other hand, regardless of the strengths or weaknesses of the connectivity of urban areas, networked driverless cars will become an important part of the city. The core areas of developed cities will operate as unmanned shared cars and driverless buses.

智能家居控制系统
SMART HOME CONTROL SYSTEM

生活功能模块：从简洁和集约出发，以集水电为集中的关键点，探究可以拆卸外出的活动单元。集合：将最基本的生活功能集合成模块，包含卫生间、厨房、冷藏、洗衣烘干和居住为一体的集合功能块，并将所有水电以及生活垃圾进行集中处理。伸缩：模块内空间可通过内置伸缩进行扩大和缩小以满足各种使用习惯。拆卸：模块中的功能与摆放位置可根据喜好进行拆卸和组装。出行：模块与无人驾驶出行相结合，形成新一代出行旅居场所。

Life function module: Starting from simplicity and intensiveness, the key points of collecting hydro power and electricity are gathered, and the active units that can be disassembled are explored. Collection: The most basic life functions are assembled into modules, which include a collection function block of the bathroom, kitchen, refrigeration, laundry, drying and living as a whole, and all the hydropower, electricity and domestic waste are centralized. Telescopic: The space inside the module can be expanded and reduced by built-in adjustability to meet various usage habits. Disassembly: The functions and placement of the module can be disassembled and assembled according to preferences. Travel: Modules are combined with unmanned travel to form a new generation of travel and accommodation.

未来城市划分
FUTURE URBAN DIVISION

随着交通工具速度和自动驾驶功能达到新的高度，交通工具的时速已经可以和现在的高铁动城相比拟，移居已经变成轻而易举的事情，未来的城市格局不再像现在根据省市直辖市进行划分，城市的功能和格局也不会像现在这样密集和复杂，更多的城市将主要赋予专一的功能，科技区，渔业区、交通运输区、林牧区、谷物区、牲畜矿业区、金融区，教育区，政治区，医疗区。也许在一个区域所包含的大大小小业态将扩大到整个国家，所有的行业以及产业将进行统一的聚集和生产。在高密度聚集生产的环境下将会引起各个行业的巨大进步和发展，整体带动各个行业的交流和传播，提高整体行业影响力。由于速度的提升，在区域内将不存在明显的地域优势，人们可以进行方便的移居。而对于公民，将少于奔波之苦，在各个城市之中可以体验接收到最高精尖的服务和知识。

With the speed of vehicles and the automatic driving function reaching new heights, the speed of vehicles can be comparable with the current high-speed rail. Migration can be achieved without any difficulty. The future of urban structures is no longer like the current municipals as it won't be divided according to provinces, cities and municipalities directly under the central government. The functions and patterns of cities will not be as dense and complex as they are currently. More cities will mainly be given specific functions, such as science and technology zones, fishery zones, transportation zones, forestry and pastoral areas, grain areas, livestock mining areas, finance areas, educational areas, political areas and medical areas. Perhaps the size of the industry included in one region will be extended to the entirecountry, and all of the industries will be unified and aggregated in production. In theenvironment of high-density agglomeration production, it will cause tremendous progress and development in various industries, and promote the communication and dissemination of various industries as a whole industry, and improve the overall industry influence. Due to the increase in speed, there will be no obvious geographical advantages in the region, and people can carry out migrate conveniently. As for citizens, they will not have to rush around as much, and they can experience the highest level of service and knowledge in each city.

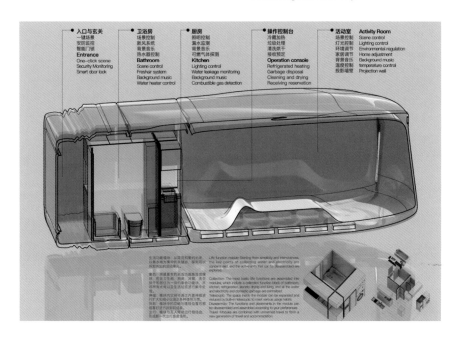

荣获奖项：入围奖　　　　　　　Awarded: Finalist Award

拉近
——智慧的垂直生活理念

Closer
——Smart Vertical Life Concept

团队成员：Sadyr Khabukhayev
来自：高句丽大学

Team members: Sadyr Khabukhayev
From: Koguryeo College

概念
CONCEPT

工业革命后，越来越多的人从乡村移居到城市。为了满足人民不断增长的居住需求，城市内修建了越来越多的建筑。然而，这些建筑却丧失了人性尺度，亦未体现自然。假设我们恢复住宅建筑中的人性尺度会怎么样？这将有利于拉近居民与大自然之间的关系。

After industrial revolution, more people moved from villages to cities. Architecture was made to meet the increasing accommodation demands, simultaneously losing human scale and presence of nature. What if we bring back human scale to residential architecture? Nature and neighbors become closer.

技术使人们远离大自然和社区是一种常见的偏见，本项目的本质是凭借而不是忽略技术来拉近人与人的距离。

The conventional preconception is that technology seems to alienate people from nature and community itself. This projects essence is to bring people closer, not in spite of technology but because of it.

新设计的图标
NEW DESIGN ICON

在视觉上,设计方案延续了现有的审美主题,并加以拓展,保持了整体的和谐。遮阳篷的棱角成为整栋建筑的象征。

Visually, the design proposal continues the existing aesthetic motif but expands it, preserving an overall harmony. The angular shape of sunshades becomes a symbol of the whole building.

新住宅楼内设置了绿色悬臂式露台,不仅拉近了人与自然的关系,也减少了热辐射,有利于调节内部的气候。Etfe/钢结构外壳内设置了悬臂式包层,功能之一是充当遮阳篷,同时也允许建筑物接受周围的自然光线的照射。悬臂式包层的另一项功能是减少风量,居民能够更加享受在露台上的时光。此外,外观设计也掩盖了建筑内部用于灌溉的集水和供水系统。

New residential floors have cantilevered green terraces which not only bring people closer to nature, but also reduce heat radiation and allow better inner climate control. Cantilevered envelope is embraced into outer shell of etfe/ steel structure, which acts as a sunshade but permits ambient natural light into the building. Another function is to break wind masses to make time being on terraces more pleasant. Moreover, the exterior conceals rainwater harvesting and water supply system for irrigation inside of it.

应用程序
APP

新住所配备了一款桌面和移动应用程序。本应用程序有两个基本的目的:

The new residencies come with an APP developed for desk top and mobile use. This APP has two basic purposes:

1. 社交
   Social

2. 智能家居
   Smart home

动能
KINETIC ENERGY

每栋楼的前两层和走廊都配备厨房地砖,将在厨房行走的人的能量转换为电力,以用于照明和机械服务等。

First two floors and corridors on every level are equipped with kitchen floor tiles which convert the kitchen energy of walking people into electricity for lighting, mechanical services, etc.

集水
RAINWATER HARVESTING

太阳能
SOLAR ENERGY

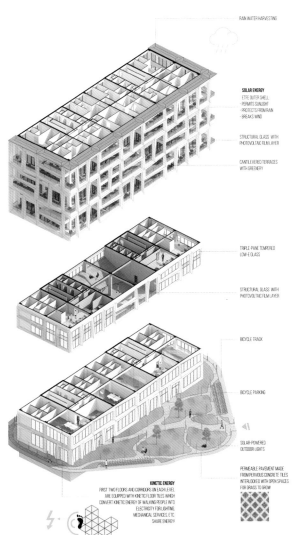

Etfe 外壳:
Etfe outer shell:
允许光照
Permits sunlight
防止雨水
Protects from rain
挡风
Breaks wind

有光伏薄膜层的结构玻璃
Structural glass with photovoltaic film layer

带有绿色植物的悬臂式露台
Cantilevered terraces with greenery

三层钢化
Triple-pane tempered

Low-E 玻璃
Low-E glass

有光伏薄膜层的结构玻璃
Structural glass with photovoltaic film layer

自行车道
Bicycle track

自行车停放处
Bicycle parking

太阳能
Solar-powered

太阳能室外灯
Outdoor lights

早先的混凝土瓦片制成的透水路面应与空地连结,供草生长。
Permeable pavement made from previous concrete tiles is interlocked with open space for grass to grow.

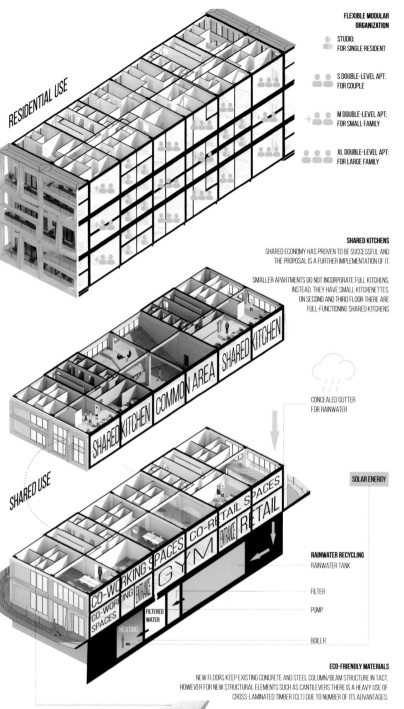

柔性模块化组织
FLEXIBLE MODULAR ORGANIZATION

单身公寓：适宜单身居民
Single apt: for single resident
双层单身公寓：适宜夫妻
Double-level single apt: for couple
中户型双层公寓：适宜小家庭
M double-level apt: for small family
超大型双层公寓：适宜大家庭
XL double-level apt: for large family

共享厨房
SHARED KITCHENS

共享经济已证明是成功的；建议进一步实施共享经济。小型公寓不包含完整的厨房。但是，它们有小型厨房。此外，小型公寓的三楼和四楼有全功能共享厨房。
Shared economy has proven to be successful and it is proposed to further implement it. Smaller apartments do not incorporate full kitchens. Instead, they have small kitchenettes. On second and third floor there are full-functions shared kitchens.

隐蔽雨水排水沟
Concealed gutter for rainwater

太阳能
SOLAR ENERGY

雨水回收
RAINWATER RECYCLING

雨水池	泵
Rainwater tank	Pump
过滤器	热水器
Filter	Boiler

生态友好型建筑材料
ECO-FRIENDLY BUILDING MATERIALS

新型建筑确保现有混凝土、钢柱/钢梁的完整性。然而，就新型结构元件例如悬臂而言，交叉复合木材 (CLT) 因其以下几点优势而大量使用。
New floors keep existing concrete and steel column/ beam structure in tact. However, for new structural elements such as cantilevers, there is a heavy use of cross-laminated timber (CLT) due to the advantages listed below:

1. 零碳足迹
1. Zero carbon footprint

2. 相比混凝土，交叉复合木材重量较轻，同时保留了抗震性能和耐火性能。
2.Less weight compared with concrete while preserving earthquake and fire resistance.

3. 预加工降低建筑成本，缩短建筑工期，减少建筑垃圾。
3.Lower construction costs, time and waste due to prefabrication.

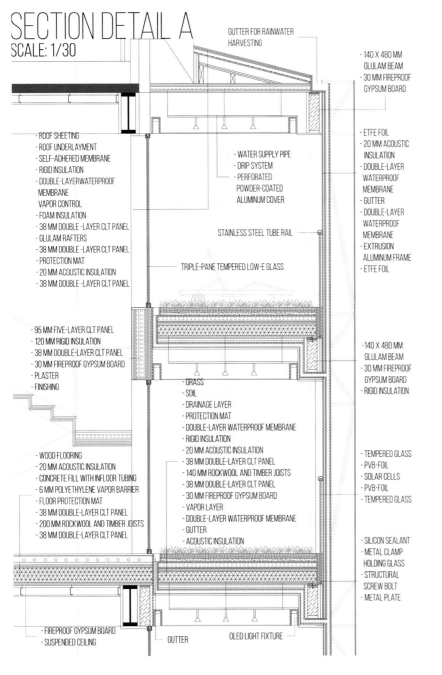

SECTION DETAIL A
SCALE: 1/30

剖面详图 A
FSECTION DETAIL A

Scale: 1/30

集水沟
Gutter for rainwater harvesting

140 毫米 ×480 毫米胶合梁
140mm×480mm glulam beam

30 毫米防火石膏板
30mm fireproof gypsum board

顶板
Roof sheeting

屋顶垫层防水膜
Roof underlayment

自粘卷材
Self-adhered membrane

刚性绝缘
Rigid insulation

双层防水膜蒸汽控制
Double-layer waterproof Membrane Vapor control

泡沫绝缘
Foam insulation

38 毫米双层交叉复合木板
38 mm double-layer CLT panel

胶合橡
Glulam rafters

38 毫米双层交叉复合木板
38 mm double -layer CLT panel

防护垫
Protection mat

20 毫米隔声设备
20 mm acoustic insulation

38 毫米双层交叉复合木板
38 mm doubler-layer CLT panel

供水管
Water supply pipe

滴灌系统
Drip system

穿孔动力涂层铝盖
Perforated power-coated aluminum cover

不锈钢轨
Stainless steel tube rail

三层钢化 low-E 玻璃
Triple-pane tempered low-E glass

Etfe 箔
Etfe foil Etfe

20 毫米隔声设备
20 mm acoustic insulation

双层防水膜
Double-layer waterproof membrane

排水沟
Gutter

双层防水膜
Double-layer waterproof membrane

挤压铝框
Extrusion aluminum frame

Etfe 箔
Etfe foil Etfe

95 毫米五层交叉复合木板
95 mm five-layer CLT panel

120 毫米刚性绝缘
120mm rigid insulation

38 毫米双层交叉复合木板
38mm double-layer CLT panel

30 毫米防火石膏板
30mm fireproof gypsum board

石膏
Plaster

终饰
Finishing

30 毫米防火石膏板
30mm fireproof gypsum board

蒸汽层
Vapor layer

双层防水膜
Double-layer waterproof membrane

排水沟
Gutter

隔声设备
Acoustic insulation

刚性绝缘
Rigid insulation

20 毫米隔声设备
20 mm acoustic insulation

38 毫米双层交叉复合木板
38mm doubler-layer CLT panel

140 毫米岩棉和木托梁
140mm rockwool and timber joists

草地
Grass

土
Soil

排水层
Drainage layer

防护垫
Protection mat

双层防水膜
Double-layer waterproof membrane

Vicente Guallart

巴塞罗那前城市总建筑师,加泰罗尼亚高等建筑学院 (IAAC) 创始人
Former Chief Architect of Barcelona Municipality
Founder of the Institute of Advanced Architecture of Catalonia (IAAC)

科技重构城市生态
Science and Technologies Restructure Urban Ecosystems

建筑产自当前,却会屹立百年。建筑师必须时刻思考未来。下一代建筑竞赛是探讨建筑未来的完美开端,下一代建筑奖是一个极具开创性的奖项。

As buildings constructed today would last for one hundred years or more, architects should always be foresighted. Next Architecture competition is a perfect start for exploration and discussion on the future of buildings and Next Architecture award is extremely pioneering.

今天,我们迈入了全新的时代,数字世界完全改变了人们的生活方式、通信和社交方式,但还未完全改变我们的生活环境。显然,数字科技也将在未来带来建筑与城市的巨变。数字技术不单会重新定义建造的方法,还将影响我们对城市的运营规划。将来,建筑是一个有机体,会在内部进行新陈代谢,并利用有效信息进行更好的居住空间管理,同时,数字技术也会重新定义建筑服务与管理。最终,数字技术将成为建筑发展的基础。

As we enter a new era, digitization has altered people's ways of living, communication and social interaction, but our living environment has not been changed yet. Obviously, digital technologies would drastically change buildings and cities in the future. Not only would digital technologies redefine the methods of construction, but they would influence the planning and operation of cities. Buildings would become organic beings with internal metabolic system. Buildings themselves could use effective information to make better management of the living space, and meanwhile, digital technologies would also redefine services and management of buildings. Eventually, digital technologies would become the basis for architectural development.

在建筑创新方面,西班牙历史悠久。然而,面对当前科技迅速发展,西班牙建筑师依然需要重新理解建筑。在西班牙,住房价值重在社会导向,建筑通常造价低廉,不追求豪华设计。这种以大众为方向解决问题的方式,决定了对被动建筑设计的关注,也意味着建筑师需要利用科技来建造更节能的建筑。另一方面,在技术行业,由于西班牙光照充足,能源的区块链技术与能源生产技术发展良好。未来的目标是整合这两方面,让更多的人享受更好的建筑和住房,同时利用数字技术为建筑提供更多的可能性。

Spain has a long history of architecture innovation. Yet against the backdrop of rapid scientific and technological development, Spanish architects should rethink about architecture. Based on people's needs, buildings in Spain usually are neither costly nor luxurious. Because of this consideration for people's need, passive building design is popular. This also means architects need to design energy-efficient buildings with the use of technologies. On top of that, Spain has abundant sunshine and well-developed energy production technologies and blockchain technologies in the energy sector. We aim to factor in these two aspects to construct buildings with higher quality for more people and add more possibilities to buildings by the use of digital technologies.

与西班牙相比，中国在过去30年出现了许多优秀的建筑，特别是在基础设施方面处于领先地位。而与这种进步相伴的问题是，很少有时间去考虑建筑的细节、质量与公共空间的利用。当前，中国的国家战略提出了建设生态文明城市的目标，注重生态环保的建筑将是未来城市发展的主流。为了实现这一目标，我们必须思考未来建筑和城市的结构，只有把城市发展与建筑科技正确结合，才能真正实现可持续的生态目标。

Meanwhile, the past three decades have seen enormous fabulous buildings put in place in China. And it's worth noting that China has taken the lead in infrastructure construction. Despite these achievements, Chinese have seldom deliberately thought about details, quality and utilization of public spaces of buildings in China. Now building eco-friendly cities is one of China's national strategies, Eco-protection would become a major focus in urban development in the future. To this end we have to contemplate how to better design the structures of buildings and cities of the future. Only by well applying architectural technologies in urban development can we really materialize sustainable development of urban ecosystems.

将科技与建筑融合，最大的挑战在于如何将各方观点融合在同一个项目中。提供技术的IT公司、建筑的设计师、保障社会住房的政府，这三方都有各自的立场和思考，重要是有能力在新的建筑要求与城市环境下结合新的科技和方法。因此，一个优秀的设计师，必须专注于建筑的整体性。也许，下一代建筑师会类似于汽车设计师，不再区分设计、工程和结构专业，而是对设计、构建与建筑管理有全面的了解。建筑教育也需面向更宽泛的领域，不能只局限在各自细分的专业内，而是对建筑师普及设计、历史、文化与经济知识，让他们的视野关注到例如能源、交通与材料等多个方面。

When we apply technologies in construction, the greatest challenge is how to factor in opinions of various parties in one project. Architects who design the building, IT companies who provide technologies and governments who need to guarantee residence of citizens have their own positions and ways of thinking. We have to apply new technologies and methods to meet new requirements for buildings in new urban environment. Thus an excellent designer has to look at buildings from a holistic perspective. Just as auto designers, architects may not be divided into designers, engineers and tonic architects in the future, and instead they would have a holistic understanding of design, construction and building management. Moreover, architectural education should not be limited to different majors, but should extensively cover more fields. Would-be architects should be instilled with knowledge of design, history, culture and economics and thus they could also pay attention to aspects including energy, transportation and materials.

"智慧树"竞赛的参赛作品中，有的尝试将绿色生态和垂直关系引入建筑，有的尝试使用某种简单的方式开发出针对这个建筑的软件，有的在寻找新的智能和新的架构挑战……这些有价值的想法，无疑是对未来建筑一次很好的切入。对于未来的建筑设计，我们目前正处在一个关键的时刻：建筑的形态与性能都至关重要，我们希望做出既造价合理又性能优良的建筑。因此，重要的是定义原型与构建原型。一旦原型成功，就能复制出更多项目。从这个视角看，重要的不是去设计，而是去创造。

Among all entries of "Smart Tree" Competition, some adopted the concepts of green ecosystem and vertical structure in designing buildings, some attempted to develop software for the buildings by simple ways and others tapped into intelligent applications and new structures ... All these valuable ideas well explore possibilities of buildings in the future. In terms of design of future buildings, we are in a crucial moment: as both the shape and performance of buildings are very important, we hope to put in place buildings with lower cost and greater performance. Thus what's important is defining and creating prototype buildings. Once prototypes are in place, we can replicate them in many projects. What we should do is not design but creation.

Hernan Diaz Alonso

南加州建筑学院（SCI-Arc）院长和首席执行官
Director and CEO of SCI-Arc

建筑是城市变革的推动者
Architecture is the Promoter of Urban Change

建筑需要代表它的时代，同时也向往永恒。
Architecture needs to not only represent its era, but pursue eternity timeless.

在未来三四十年内，我们面对的问题是：城市人口大量增加，密度依然是影响可持续性与生态环境最至关重要的问题。建筑需要找到维系城市运营及高效利用能源的最佳方式。很可惜，我们创新的速度依然没达到的标准，无论对于开发商还是居住者，我们还困在传统的观念之中，很难解决城市住房问题。
In the next 30 to 40 years, the population in cities is going to increase exponentially. Density is a crucial problem affecting sustainability and the ecological environment. Buildings need to find the best way to sustain urban operations and use energy efficiently. Unfortunately, we're not moving as fast as we should and innovation is not yet up to the standard. To developers or occupants alike, we are stuck in traditional concepts and can hardly solve housing problems.

此次竞赛，聚焦未来城市的发展，也意味着我们面对困境，去探讨得出更创新的思考方式。而面对不同的裁判员、不同的变量、不同的问题以及可实施的事情，一些最激进的想法也意味着积极变革的第一步。我坚信建筑有能力以许多其他领域无法做到的方式，将社会意愿综合在一起。建筑应该成为城市变革的推动者。
As this competition focuses on the development of future cities, we must explore more innovative ways of thinking as we face difficulties. In the face of different referees, variables, problems and things that can be implemented, some most radical ideas also mean the first step towards positive changes. I am convinced that architecture has the capacity to synthesize societal desires and societal aspirations in a way that many other fields cannot do. Architecture ought to be the driving force of urban change.

首先，建筑必须变成一个平台，让各种技术可以在上面发挥效力。由于很多科技根本就不需要建筑设计的物理存在，而且建筑不够快，无法吸收每一项科学技术，所以，建筑设计需要不断地重新创造自己对科技的理解，并使自身成为一个能吸收最新科技力量的平台。科技是多层次的，建筑设计能在很多方面与科技相联系，比如界面交互、社交媒体、通信技术，人工智能等等。实际上，比起建筑，设计与现代科技融合地更好，因为建筑比设计体量更大、时间更长、花费更多。建造的整个环节可能需要 3 到 5 年的时间，而 5 年内，科技会发生很多变化，也正因此，建筑设计需要用一个更广阔的视角来考虑科技的定位，而不能成为技术即时性的仆人。

Architecture must become a platform in which technologies can be applied. As the latest technology doesn't really require the physical existence of architectural design and architecture is not fast enough to absorb every aspect of the latest technology; architectural design has to constantly recreate itself and make themselves a platform in order to absorb and understand the role of the latest technologies. There are multiple layers in technology, and architectural design can relate to it in an amplitude of ways, such as interface interaction, social media, communication technology, artificial intelligence and so on. In fact, the design aspect is very much in tune with contemporary technologies when compared with architecture for it is bigger in size, more time consuming and costly. The process of building between design and construction may take 3 to 5 years, and in those 5 years the change in technology is vast. Therefore architectural design has to consider the positioning of technology from a broader view, and we cannot become servants of the immediacy of technology.

对比科技，建筑更为持久。建筑是城市文化的代言人。它不仅反映了当下的科技与时代，也塑造文化，塑造我们看待和感受城市的方式。实际上，关于高密度住房的讨论和建议，通常总会太专注于服务和功能，不够关注建筑美学和整体城市美学。如果在这两者之间进行平衡，我希望建筑成为一个具有包容性的容器，既服务于技术与功能，也确保设计的力量使审美始终存在。

Compared to technology, architecture is the spokesperson of urban culture as it is very durable. Architecture not only reflects the current technology but also shapes culture and how we see and experience the cities. In fact, more often than not, some of these discussions and suggestions in relation to high-density housing tend to focus on service and functional parts too much, and not enough on architectural aesthetics and overall urban aesthetics.To achieve a balance between the both, I hope that architecture will become an inclusive container which serves both technology and function while ensuring that the power of design keeps aesthetics in existence.

数字技术的发展将会给建筑语言、建筑师带来新的变革，计算工具也会催生出全新的几何构造。传统建筑将拥抱新的技术，计算机也会用更加理性的方式来强调美。来自影像、拓扑形态、流行文化等形式的传统元素可以成为当代数字技术下自由形式的基础。将极大扩展建筑美学的力量，挑战对建筑既有的认知。建筑代表设计的力量，而技术让我们传达这样的力量。如果说传统的建筑学是在用已有的材料和技术探索建筑美的极致，那么未来建筑，能让我们多一种看待世界的方式。建筑是生动的且适当无用的。建筑的灵魂在于想象并全方面打破文化的教条。我所期待的城市未来，也是一种能同时推动美学、效率、品质以及可持续性的创造。

As digital technology develops, it will bring new adaptations to architectural language and architects. Computing tools will also create new geometric structures. The traditional form of architecture will embrace the latest technologies as well as emphasize physical aesthetics in a more rational way through the use of computers. Elements such as images, topologies and popular culture can become the basis for free form in the contemporary era of digital technology. In turn, it will greatly develop the ability of architectural aesthetics and challenge the existing understanding of architecture. Furthermore, architecture is representative of the power of design and technology enables us to convey such power. If traditional architecture taps into the ultimate beauty in buildings using existing materials and technologies, architecture of the next generation would shed new light on our world view. Architecture could be vivid and appropriately non-utilitarian. The soul of architecture lies in imagination and breaking away from doctrine. I also hope that future city will become a system of sustainable innovations with aesthetics, efficiency and quality.

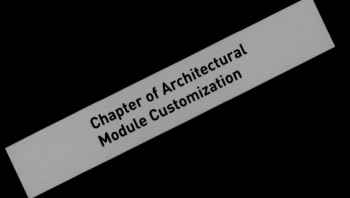

Chapter of Architectural Module Customization

建筑模块定制篇

荣获奖项：模块奖　　Awarded: Unit Module Prize

未来垂直村
——AI 匹配的装配式自进化栖居

The Future of Vertical Village
—— AI Matched Fabricated Self-Evolving Residence

团队成员：陆超，翁申霞（个人）
来自：华南理工大学建筑设计研究院

Team members: Lu Chao, Weng Shenxia (Individual)
From: Institute of Architectural Design, South China University of Technology

在第七届中德经济技术合作论坛上，李克强总理把一款精巧的鲁班锁送给德国总理默克尔。一个小小的鲁班锁，其实蕴含着中华传统的大智慧。我们在想，未来的居住单元必然可以进行个性化定制，人们可以在 APP 里选择自己的需求，得出最合适的户型。然后这些户型模块可以像鲁班锁一样，通过精巧的扣合，成为一座大厦的整体。

At the 7th Sino-German Economic and Technical Cooperation Forum, Premier of the State Council Li Keqiang presented an exquisite Luban lock to German Chancellor Angela Merkel. As a matter of fact, a small Luban lock encompasses the great wisdom of Chinese tradition. We are thinking that personalized customization is inevitable for the future residential units. People can choose their own needs within the APP, and can get the most suitable type of housing. Then these household modules can act like the Luban lock, becoming part of the entire building through the ingenious combination of interlocking.

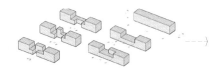

以此，我们提出了未来栖居的方案：
In this way, we put forward the plan of future residential living:

我们设想了一个长 8 米，宽 3 米，高 3.3 米的基本箱体单元，并由这个基本箱体单元衍生出 12 种户型。这些户型在工厂里预装好各种现代生活设施，然后通过装配式技术，装嵌到大厦的梁柱结构中。
Our assumption is a basic box unit of 8 meters long, 3 meters wide and 3.3 meters high, which derives from 12 types of households from this basic box unit. These apartments are preassembled in the factory with various modern living facilities, and then they are embedded into the beam-column structure of the building through assembly technology.

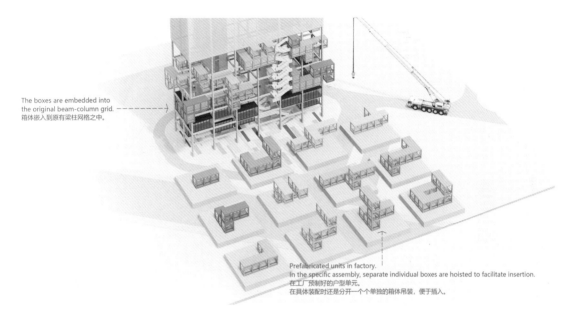

这个过程有点像鲁班锁的扣合，是一种非常有趣的咬合、缠绕。而且奇妙的是，我们可以不刻意把它填满，而是让它自然地留有一些空位，作为公共的绿化、交往平台。
This process is a bit like the interlocking of a Luban lock. It has a very interesting occlusion and bind. And what is amazing, is that we do not deliberately fill it to the full, but let it naturally leave some unoccupied space as a communal green area anda communication roof terrace.

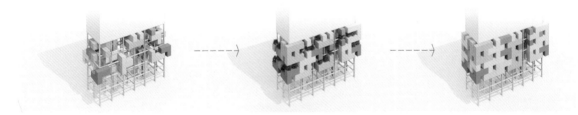

而低层的空间，也采用装配式的箱体，提供超市、餐饮、咖啡、健身、电影、阅览、游艺、会议、温室农场等各种功能模块，提供公共服务空间。
The lower-level space also uses assembled boxes to provide supermarkets, catering, coffee, fitness, movies, reading, entertainment, meetings, greenhouse farms and other functional modules to provide communal service space.

当然，一道炫酷的室外捷径楼梯也是增色不少的。
这样，就成了一个拥有众多空中绿化与交往空间的"未来垂直村"。
Of course, a cool outdoor staircase allowing shortcuts can also add a lot of fun.
In this way, it has become a "future vertical village" with abundaspace for aerial greening and communication.

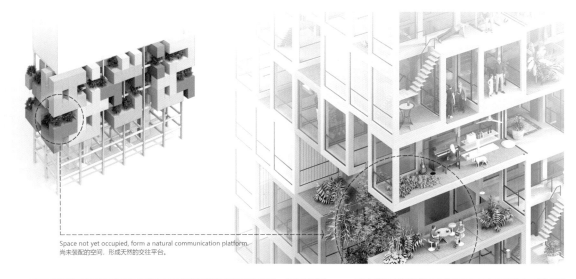

Space not yet occupied, form a natural communication platform.
尚未装配的空间，形成天然的交往平台。

尚未装配箱体的空间，就成了天然的交往平台。
一些箱体的缺失，为创造更多的交往提供了机会。

The space that has not yet been assembled naturally become a social roof terrace.
The absence of some boxes provides opportunities for creating more social interactions.

Traditional "side-by-side" apartment usually has only two neighbors.
传统的"并排式"户型，一般只有两个邻居。

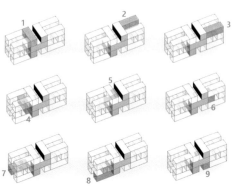

— 52 —

传统的住宅，一般只有左、右两个邻居。而创新的"鲁班锁式"户型，因为跨越了不同的楼层，每户可以紧邻多达9个邻居，大大提升了交往的可能性。

Traditional houses generally only have two neighbors, to the both sides. The innovative "Luban Lock" household, spans across different floors, therefore each household can be adjacent to up to 9 neighbors, greatly enhancing the likelihood of social interaction.

而即使是箱体装满的状态，每个箱体之间还是留有"共享小阳台"的：下图是假想的黄先生的家，是一个跨越了三层楼的户型。这天，黄先生在自家三楼的公共小阳台浇花的时候，遇到了邻居紫先生，畅聊了起来。而在这个仅设置了六层的小小"垂直村"，这样的公共小阳台就多达126个！这些愉快的交往每天都在不同的角落发生。例如黄先生的家，就共接触到6个共享小阳台。这些小阳台上的植物，都是黄先生与各个不同的邻居共同打理的。融洽的邻里关系也由此发展。

And even if the state of the boxes is full, there is still a "shared balcony" between each box: The following picture is a hypothetical image of Mr. Huang's home, which spans across three floors. On this day, when Mr. Huang was on the third floor of his house watering the flowers on the communal, he met Mr. Zi, a neighbor, and chatted happily. In this small "vertical village" with only six floors, there are as many as 126 communal balconies. These pleasant interactions may take place in different corners every day. For example, Mr. Huang's home has six shared balconies. The plants on these small balconies are taken care of by Mr. Huang and his different neighbors. From this, harmonious neighborhood relationships have developed.

这些户型的组合，其实并没有固定的式样，而是可以像搭积木那样千变万化。因此，最好就是有一个AI（人工智能），来匹配各个住户，根据他们的兴趣爱好、生活需求，安排到一起，然后又刚好组合出一个大厦的整体。

The assembly of these units, in fact, does not have a fixed pattern, but can be just as interchangeable as building blocks. Therefore, it is best to utilize AI (Artificial Intelligence) to match the various households, according to their interests, life needs,and then just integrate the households into the building as a whole.

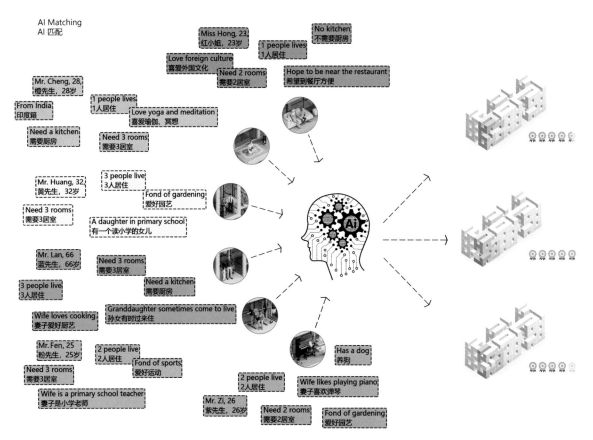

Suppose there is an artificial intelligence program that can generate the most reasonable combination of spaces according to the needs of the occupants.
设想有一个人工智能的程序，可以根据住户的需求，生成最合理的空间组合。

现在人工智能飞速发展，已经可以作曲、绘画、下棋……
这些户型的组合，其实跟作曲很相似。因此，我们用中国的"传统十二音律"来命名了十二种单元模块户型。这些户型，将会通过人工智能，匹配出最美的曲调。

Nowadays with the rapid development of artificial intelligence, you can already have the ability to compose music, paint and play chess… The assembly of these units is very similar to composing music. Therefore, we use the Chinese "traditional twelve tone row" to name the twelve unit modules. These units will match the most beautiful melodies through the use of artificial intelligence.

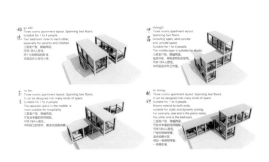

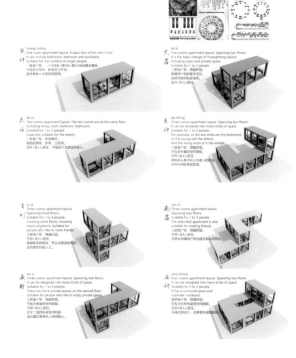

1. Ancient times, world of dinosaurs
1. 远古时代，恐龙的世界
2. Mankind grows and rules the earth.
2. 人类壮大，统治地球
3. Human retreats and other forms of life become strong. As one falls, another rises.
3. 人类退居，别的生命形式壮大。此消彼长。

更激动人心的是，这个居住系统，竟然还能自己"进化"。

由于组合、拆装方便，整个居住群落将处于不断的发展之中。就像地球上曾经发生的那样：

在初期，住户不多，剩余的交往空间就很多。渐渐地，更多住户搬入，公共空间也会变少。于是，又有住户会搬到附近新的大厦去，腾出位置。而有钱的住户，看到邻居搬走，也可以赶快购置更多的箱体模块，装到自家旁边，扩展居住空间……

What is more exciting is that the residential system can unexpectedly evolve itself.

Because of the convenience of mounting, disassembly and assembly, the whole residential community will be in continuous development. Just like what once happened on Earth:

In the initial stage, there were not many households and there was a lot of space left for social interactions. Gradually, more residents will move in and less communal space will be available. As a result, other tenants will vacate to a new building nearby to make more room. Wealthy residents, after seeing their neighbors move away, can also quickly buy more box modules and put them next to their homes to expand their living space...

整个居住群落处于不断地更新变化之中，不断达到新的平衡。

当然，箱体自身也是随着科技不断进化的。可以整体维修、更换、搬迁。搬家再也不用收拾行李啦！整个箱体拆装就可以了。

因为集合了以上这么多优点，所以我们的作品名字有点长，叫做"AI匹配的装配式自进化栖居"。

The entire residential community is in constantly renewal and change, continuously reaching a new balance.

Of course, the box itself is continuously evolving with technology. It can be repaired, replaced and relocated as a whole. You no longer need to pack to relocate! The whole box can be disassembled and reassembled.

Because of all these advantages, our work has a long name, called "AI Matched Fabricated Self-Evolving Residence".

荣获奖项：模块奖　　　　Awarded: Unit Module Prize

WONDERLAND
——传统生活的新回归

WONDERLAND
——The New Return of Traditional Living

团队成员：王涛 王智吏 陈佳 高龙起 蔡江豪
来自：安宸建筑设计咨询（上海）有限公司

Team members: Wang Tao, Wang Zhili, Chen Jia, Gao Longqi, Cai Jianghao
From: ANT Architectural Design Consulting (Shanghai) Co., Ltd.

记忆中，去邻居家"串门"是件有趣的事儿。一个普通的午后，家家都敞开着大门，有人在二楼卧室打盹，有小孩在门前跳绳。邻居正切了西瓜，一块块送到左邻右舍，顺道坐下一起红白机对战一回合。而现在，有生鲜一小时送达服务，可以只买几瓣西瓜，新鲜切好还包甜。生活是更便利了，但似乎失去了一点人情味。邻里间，多是点头之交，还是因噪声矛盾认识的；独居老人，去世很久，才能被发现；小孩，为了安全，常被锁在家中，孤独地望着窗外。

In the olden times, dropping at the neighbor's is such a fun. All the houses in the little community were opened to the neighbors on an ordinary afternoon. Doors open, one may take a nap in the bedroom on the second floor; the children were having fun with rope skipping; one of the neighbors is having a watermelon cut, and her child was dispatched for sharing, on whose way round, he saw other children playing with game machine, and joined in. However, nowadays, fresh food can be delivered in a one-hour service. You can buy several pieces of watermelon, which is fresh and sweet. Life is more convenient, but with less human interaction. Trapped inside concrete tower blocks, neighbors are barely interacting with each other, and some of the limited communications are created by noise dispute; elders are found having dead for long; children are locked inside and can only looking outside the window.

千百年来，家与其周边的人和环境之间的互动成为社会生活的重要组成部分，而现在，科技让人们的交流越来越间接，现代建筑已经极大地削弱了这种互动的机会。家变得更加孤立。人们越来越乐于生活在网络的虚拟空间之中，我们是想被科技一步步带进"骇客帝国"的世界？还是想引领科技创新更美好的现实生活？

From a long time ago, communication within the neighbourhood has always been an important aspect of social life. But now, technology has allowed people to communicate easily through indirect methods. Modern architecture has given fewer opportunities for people to interact. Homes have become isolated and people are more and more willing to live in the virtual space. Do we want to gradually go into the "Matrix" world in the future? Or do we want to live in a better real world by using technological innovations?

我们的设计便源于此。我们思考，如何在科技进步的历史长河大背景下，在未来建筑中，创造类似过去的邻里关系？

This is our original thought and reason behind our design. We are thinking about how to create good relationships between neighbors like the past within future architecture under the background of scientific and technological progress?

家——邻里——社群，模块系统
HOME——NEIGHBORHOOD——COMMUNITY, MODULE SYSTEM

家——模块化，科技化但不失温馨和个性：
HOME——MODULE. SCIENCE AND TECHNOLOGICAL TECHNIQUE BUT WITHOUT LOSING WARMTH AND INDIVIDUALITY:

我们设计的家，是有私人模块和共享模块组成的。私人模块，由6个子模块组成，分别是卧室、柜橱、卫生间、DIY 模块和两块走道模块。DIY 模块可以是厨房、小客厅、创意工作室等。还可以将两户组合，连接形成一个更大的组合公寓，来满足稍大规模家庭的生活需求。

All of the "Smart Tree Village" Apartments are built up with "Private Modules" and "Sharing Modules". The Private Module is built up with six Sub Private Modules: a Closet Module, a Bedroom module, a Toilet Module, a DIY Module and two Corridor Modules. The DIY Module has several alternatives like a kitchen, a mini living room, creative studio and so on. A Double Living Module is built up with two basic Living Modules to form a larger apartment to meet the needs of a larger family.

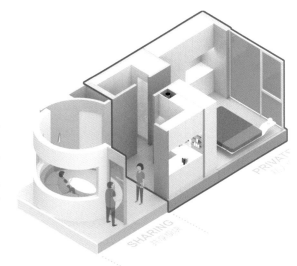

  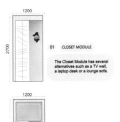 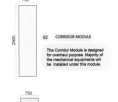

我们的重头戏，共享模块，是智慧树社区的一大特点，每一个居住模块必须有一个共享模块，这些共享模块除了居住模块的主人自己使用，也可分享给同社区邻里使用，模块类型多种多样，如共享客厅、共享办公室、共享会议室、共享微型电影院、共享健身房、共享儿童室、共享菜园、共享花园等等，你还可以根据需求定制自己的共享模块。

Our key highlight is the SHARING MODULE which is the major feature of the 'Smart Tree' vertical community. As one of the major features, each apartment, as called Residential Module, has a Private and a Sharing Module. Sharing Modules could be used only by the householders or shared with neighbors. There are different kinds of Sharing Modules ready to be chosen such as the sharing living room, sharing office, sharing meeting room, sharing mini cinema, sharing gymnasium, sharing children room, sharing vegetable room, and sharing garden, etc. The Sharing Module also could be customized according to different personal needs.

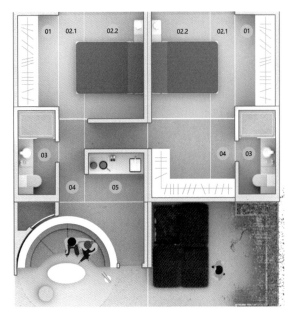

这个共享空间是鼓励居民们的积极正面的交流，学习、分享、互助友爱。除了智能化的监管机制外，基于区块链的共享积分会被应用到其中，共享模块的开放和被使用都会获得积分，反之会消费积分。

The Sharing Modules aim to increase the positive interaction among neighbors through learning, sharing, mutual help and friendship. Except for the iHome system, the Sharing Credits based on the Block-chain concept will be used widely inside the 'Smart Tree' Village. As a householder, the Sharing Credits will be earned by lending out the Sharing Module and the Sharing Credits cost for borrowing other neighbor's Sharing Modules.

共享模块中很重要的一类是共享办公室，提供了一个离家近但又独立的办公环境，它可以是一个三面显示屏的房间，可以模拟工位区，模拟会议室，虚拟现实的环境就如在公司工作一般。

One of the most valuable types of sharing modules is the sharing office. A Virtual Office provides the best office environment just right beside your home yet still independent. A virtual Office could install 3 screens on 3 walls to simulate a realistic office environment, work-station area or conference room.

邻里——资源共享，关系紧密但独立：
NEIGHBORHOOD——RESOURCE-SHARING, CLOSE RELATIONSHIPS BUT ALSO INDEPENDENT

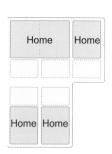

这样相近的三五家，就形成了邻里的概念。邻里紧密但独立，现在大家庭中的老人和子女通常住得比较远，但邻里可以互助。大家庭也可以选择住在隔壁，但又保证相互的独立，解决了大家庭矛盾的问题。

Three to five homes form a neighborhood. Neighborhoods are close but also independent. Nowadays, older generations and their children often live far away from each other, but neighbors can help each other. Big families can also choose to live as neighbors. In the 'Smart Tree' design, people can also live in independence in order to reduce big family conflicts.

社群——物以类聚，人以群分，慢慢形成社群文化：
COMMUNITIES——BIRDS OF A FEATHER FLOCK TOGETHER, FORMING A COMMUNITY CULTURE:

Public Platform: open space, rentable

Public Rooms: Long-term lease to merchants and operators

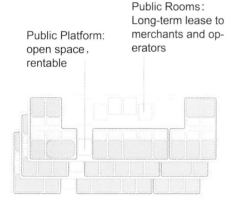

3 到 10 个邻里，形成社群。除了上述的居住模块，还有公共模块和开敞模块。公共模块是为社区提供便利服务的房间，如小超市、理发店、书吧、餐厅、咖啡厅、创客空间、快递收发站等等。

Three to ten neighbors can form a community. In addition to the residential modules mentioned, we provide public modules and open modules. The PUBLIC MODULE has

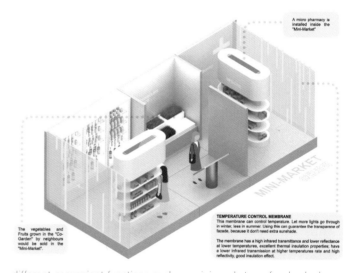

different convenient functions such as mini markets, cafes, book shops, canteens, café work studios, delivery stations and so on.

现在的高层公寓，白天只有老人和孩子，公共空间存在白天利用率低的问题。而在智慧树社区，资源将合理利用，楼内除了共享模块中的小型虚拟共享办公外，还在公共模块中提供了一些公共办公室。

Nowadays, the public area of a high-rise residential building is mainly used by children and elderly in the daytime. However, in the 'Smart Tree' village, public areas will be re-arranged to achieve the best resource allocation and solve the problem of low daytime utilization. Apart from the small virtual share office in the Sharing Module, there are also some public offices in the PUBLIC MODULE.

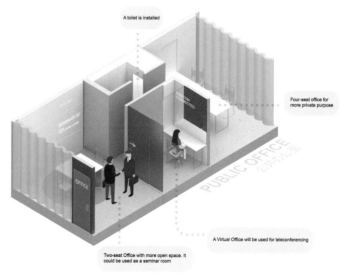

开敞模块为社区提供公共开放场所，如花园、菜园、城市农场、戏水池等等。
The OPEN MODULE will provide an open and public space for the residents in the community, such as gardens, vegetable plots, urban farmlands, swimming pools, etc.

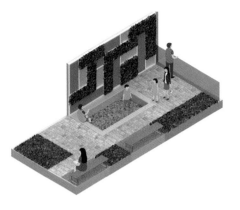

不同社群会有各自不同的特点。比如，学习社群是图书馆、书吧、书店功能多。也越来越吸引更多这样的人群，促使学习功能越来越丰富，这样形成了社群文化。
Different communities have different features. For example, learning communities have many features such as libraries, book bars and book stores. Then more and more people would like to join this community, which leads to developing an abundance of learning functions, thus a learning culture is created.

经济
THE ECONOMY

看到这，你可能会有个疑问，我们将传统的居住空间一部分分出去作为共享模块公共模块，是不是亏了。我们来算笔账。私人模块是卖的，共享模块是租的。如果你买普通一居公寓，买 34m²，使用 34m²。而在我们设计的智慧树公寓，买 21m²，租 13m²，能使用 86m²+ 的邻里共享空间以及更大的其他共享资源空间。

Now, your main concern may be whether we are losing a part of our traditional residential space. Well, let's settle it once and for all. The private module is for sale, and the shared module is for rent. If it is a normal apartment, you buy 34m², use 34m². In the 'Smart Tree' Apartment, you can buy 21m², rent 13m², use 86m²+ share space and it results in a much bigger shared space and more shared resources.

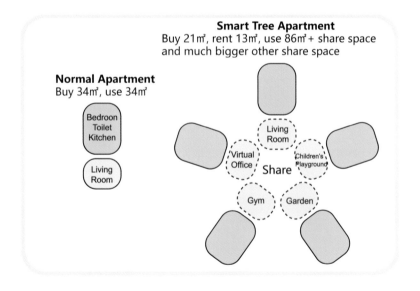

技术
TECHNIQUE

我们的设计使用了 iHome 人工智能索引系统、参数化设计、三恒系统、装配式、区块链、积分制等技术系统。非常感谢在设计过程中得到人居环境、装配式、区块链等领域的先进企业和个人的支持和指导，交流产生创造力，你们开拓了我们的思维，也提供了可实施性的依据。如故事中所说，希望在不久的将来人类引导并利用科技回归到一种更加舒适、更多交流，乐观积极的人居生活之中。

In our design, we use the iHome artificial intelligence index system, parametric design calculation, health comfort customization system, prefabricated building, block-chain, credits and so-on. We are very grateful for the support and guidance in the design process from advanced enterprises in the fields of prefabricated settlements, assembly, Block-chain areas and so on. From the support, not only have we had the technical back-up for our design, but enlarged our own knowledge reserves. We hope the human lives in the future in a way just like the story shows us: more comfortable, healthier, more communication which offers a positive residential settlement.

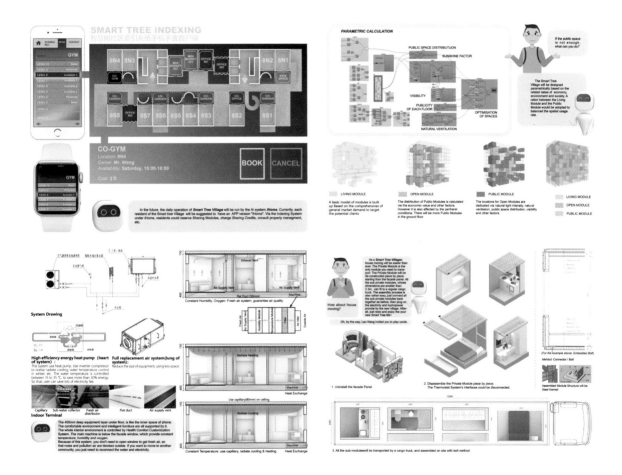

最后，希望通过此未来建筑设计，提供一种未来生活的可能性，引发大家对未来生活方式的思考，提升生活品质，对社区生活、人和人之间关系的未来发展起到正面的影响。

At last, we wish to provide a possibility of the future life through architectural design. So that people can think more about their future life style, to promote a high quality of living which may have a positive impact on community life and the relationship between neighbors.

| 荣获奖项：入围奖 | Awarded: Finalist Award |

垂直聚落　　　　　　　　Vertical Cluster

团队成员：郜佩君，段一行，徐海闻，董阳，陈乐琳，吕颖洁，刘恋
来自：东南大学

Team members: Gao Peijun, Duan Yixing, Xu Haiwen, Dong Yang, Chen Lelin, Lu Yingjie, Liu Lian
From: Southeast University

设计愿景
DESIGN VISION

基地紧邻老山国家森林公园、佛手湖公园，景观资源优渥。南京自古是山水城林相依的城市，山水、风景区成为文人雅士、乡亲贤达聚会的公共场所以及精神的寄托。经济、生态、融洽、多样的未来公寓，传统与现代、园林与城市、公共与私密的有机协同形成高效的系统。

The base is close to Laoshan National Forest Park and Buddha Lake Park. The landscape resources are excellent. Since ancient times, Nanjing has been a city adjoining mountains, rivers and forests. Landscape and scenic spots have become public places and carrier of spiritual cultivation for scholars and literati, hometown friends and relatives.Economic, ecological, harmony and diversified future apartment; traditional and modern, landscape and urban, public and private organically cooperate to form an efficient system.

因此，提出未来可定制的公寓—垂直聚落概念：
Thus, future customized apartment--Vertical Cluster is proposed：

模块建构：按照标准化模块设计，由公寓部分与社区部分构成垂直聚落单体。
Structure construction of housing modules: According to the standardized module design, the apartment and community constitute the vertical settlement unit.

自由单元
FREE UNIT

居住单元模块 Living Unit Module
社区单元模块 Community Unit Module
模块外空间 External Space of Module

垂直向公共空间 Vertical Public Space
垂直向功能服务 Vertical Infrastructure
垂直向绿色廊道 Vertical Green Corridor
垂直向社区系统 Vertical Community

使用预制装配技术，根据不同模数组合，为公寓定制提供多样的选择，满足个性化需求，经济、快速、节能、自由。单身小户型与大户型共享大户型的餐厅、花园阳台。共享空间还有小户型与小户型、中等户型之间共享的图书室、健身房、有机农场等等，居民可以通过手机终端得知共享空间的使用情况，并到达自己想到达的地方。

Using prefabricated assembly technology, we can provide various options for apartment customization according to different module combination, to meet personalized needs while being economical, fast, energy-saving and free. Dining-room and garden balcony of big household can be shared by single-small household and big household. There are also libraries, gyms, organic farms, etc. that are shared by small and medium-sized households. Residents can learn about the use of shared space through mobile terminals and reach anywhere they want.

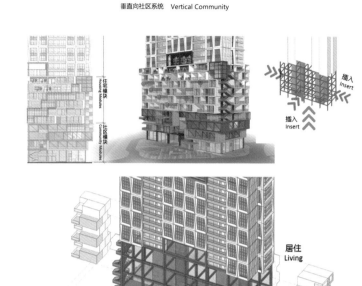

自由空间
FREE SPACE

模块外绿廊 Green Corridor
交通模块 Transport Module
模块流线 Module Streamline

预制交通模块，各类定制模块的外空间进行连接，形成吸引人集聚的趣味公共空间，提升人们自发产生交流的机会。

The external space of prefabricated traffic modules and all kinds of customized modules are connected to form interesting public space attracting people to gather and improve people's opportunities of spontaneous communication.

能源分析
ENERGY ANALYSIS

智能共享有机农场 Intelligent Shared Organic Farm

有机农场解决公寓内居民粮食需求并产生能源，自给自足，居民可通过手机终端控制灌溉农作物。

Organic farms meet the food needs of the residents in the apartment and generate energy. They are self-sufficient. And residents can control the irrigation of crops through mobile phone terminals.

智能家居通过物联网技术将家中的各种设备（如音视频设备、照明系统、窗帘控制、空调控制、安防系统、数字影院系统、网络家电以及三表抄送等）连接到一起，提供家电控制、照明控制、窗帘控制、电话远程控制、室内外遥控、防盗报警、环境监测、暖通控制、红外转发以及可编程定时控制等多种功能和手段。

Intelligent homes connects various devices in the home (such as audio and video equipment, lighting systems, curtain control, air-conditioning control, security systems, digital theater systems, Internet home appliances, and three-meter copying) through Internet of Things technology, so as to provide home appliance control, lighting control, curtain control, telephone remote control, indoor and outdoor remote control, anti-theft alarm, environmental monitoring, heating control, infrared forwarding and programmable timing control, and other functions and means.

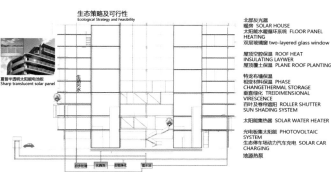

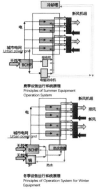
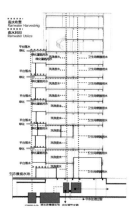

BIM 技术
BIM TECHNOLOGY

根据新型体系的应用方案，通过 Revit 软件建立 BIM 模型，考虑结构构件的预制特点以及围护体系的构造特点，选取如图所示分层装配化施工流程。

According to the application plan of the new system, the BIM model is established through Revit software. Considering the prefabrication characteristics of structural components and the construction characteristics of the enclosure system, the layered assembling construction process is selected as shown in the following figure.

荣获奖项：入围奖　　Awarded: Finalist Award

4C 生活：可持续性互联集装箱社区

4C LIFE: The Container Community of Sustainability and Interconnection

团队成员：陈嘉晖，赵佳慧
来自：华南理工大学

Team members: Chen Jiahui, Zhao Jiahui
From: South China University of Technology

用集装箱建造
CONSTRUCTION WITH CONTAINER

项目用地位于南京江北新区核心区域。东南方向 3.5km 即为中华民族的母亲河——长江。长江两岸有数个大型货运码头，最近一个距离场地仅 3km。码头存在大量可供改造的二手集装箱，为设计及建造提供了良好的基础。

江北新区人才公寓项目针对入住人群具有流动性、租赁性的特征。用集装箱进行建筑改造在价格低廉之外还满足了设计的各类需求：**模块化、标准化、可持续性、可定制化以及一定的可变性。**

The project is located at the core region of Nanjing Jiangbei New Area. 3.5 km to the southeast is the Yangtze River, the mother river of the Chinese Nation. Several large cargo terminals are situated on both sides of the Yangtze River, while the nearest one is only 3km away from the site. The used container available in large quantities lays a good foundation for design and construction.

For that the population living here are charactered by mobility and the housing here is of leasing nature, Jiangbei New Area talent apartment project is to put the container into construction, to reduce the cost as well as satisfy diverse needs during the design: **Modularization, Standardization, Sustainability, Customizability and Flexibility.**

用易得的集装箱建造
Build with easily accessible containers

连接 – 多种创意模块
CONNECTION- MULTIPLE CREATIVE MODULES

20 世纪是工业化时代，21 世纪是互联网时代，如何承接工业化时代与互联网时代，提出针对未来的居住和生活理念？物联网 + 信息互联是我提出的解决方案。我为此设计了多种创意模块，它们易于定制，便于移动，可供更换，为未来生活而设计，又根据人的意愿不断成长更新，是新一代"新陈代谢"建筑。

As we know, the 21st Century belongs to the Internet Age, while the 20th Century embraces the Industrial Age. Here comes the question: how to connect between the Internet Age and the Industrial Age and bring about the proposal which is aimed at the residence and life concept? IoT+information interconnection is the solution I raised. For that, I came up with multiple creative modules, which are apt to customize, move and change. In other words, they are the new generation buildings with metabolism.

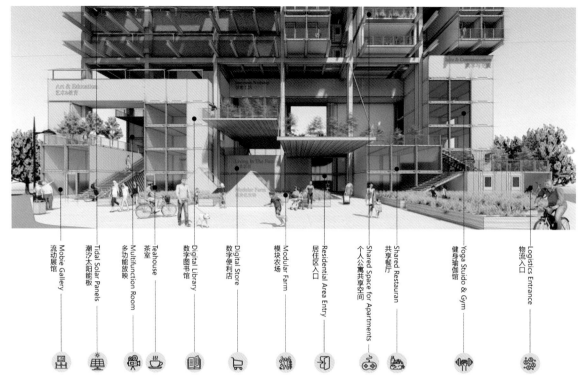

多种创意模块
Multiple Creative Modules

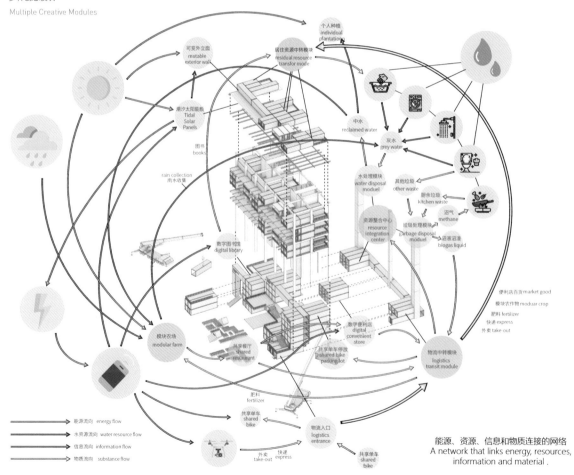

能源、资源、信息和物质连接的网络
A network that links energy, resources, information and material.

同时这些模块又不是相互独立的，它们通过信息、能量、物体而相互连接。
At the same time, the modules are not independent from each other and they can be interconnected through information, energy and objects.

可定制居住模块
CUSTOMIZABLE RESIDENCE MODULE

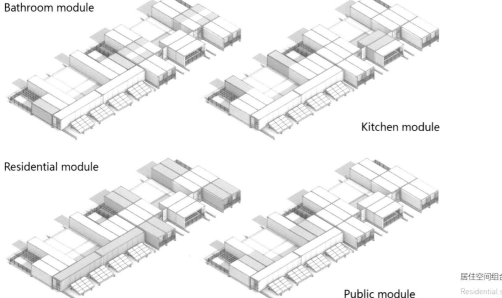

居住空间组合方式
Residential space combination

居住房型分为家庭公寓及个人公寓。得益于集装箱的丰富组合，家庭公寓中卫浴模块、餐厨模块、客厅模块、卧室模块都为可定制。个人公寓整体可定制。

The room type contains family apartment and individual apartment. What makes it special is that in the family apartment, all the modules, including bathroom module, kitchen module, living room module and bedroom module, can be customized respectively, while in the individual apartment, overall customization is provided.

居住模块通过人工智能自动判定住户是否出门。出门时收起居住可变外墙，架起太阳能板，充分利用非居住时间的泰阳能源。门厅判断住户进入建筑便收起太阳能板，打开可变外墙，引进阳光。

The Residence Module will make a judgement of whether the resident has come out automatically according to AI. When residents leave their home, the module will take down the mobile external wall and set up solar panel to make the full use of solar energy in non-residential time. Whereas the solar panel will be pulled back to let the external wall brings sunlight when AI finds out that the residents come back home.

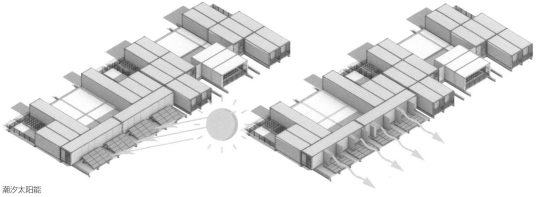

潮汐太阳能
Tidal Solar Energy

荣获奖项：入围奖　　　　Awarded: Finalist Award

智慧树
——未来垂直生活社区

SMART TREE: Future Life in a Vertical Community

团队成员：陈嘉晖，赵佳慧
来自：华南理工大学

Team members: Chen Jiahui, Zhao Jiahui
From: South China University of Technology

4.0时代的住宅设计必须做到以更智能的方式将技术融合进来。当前住宅设计已经具备同一性，出于营利目的，这种同一性主要着眼于如何创造更为密集的居住空间，却缺乏了对居住者的情感关怀，给社会群体造成了巨大的创伤。

Moving toward 4.0 era, the residential design must serve the need to integrate with technology in a smart way. The contemporary residential design has been evolved to the point of identicality which focuses on mainly density for-profit purpose, therefore, lessening emotional care for occupants leaving a rather big wound for society.

设计团队旨在关注人们的感受、需求以及进一步的期望，从而让内部和外部之间的界限不再那么泾渭分明。同时也实现了在技术应用的参与下，各个分开的新功能统一为整体的智慧建筑，并最终使人们的利益最大化。

The design Team aims to focus on what people feel, what people want and what people could expect further. Then the boundary between inside and outside is blurred out, at the same time, by the application of technologies, new functions that are separated are integrated into one, so as to benefit the residents to the largest extent.

该机制通过现有结构进行运作，让该主体结构内具有更大的空间移动灵活性，以满足人们在这个特殊时代的要求。

The Mechanism is operated through the existing structure with a more flexibility of movement of spaces within this main structure to respond to the need of people at specific times.

 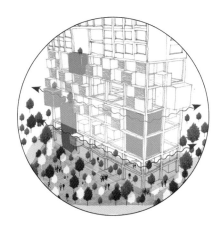

灵感
BIG IDEA

灵感源于树的三个主要部分。我们在结构中制作了不同尺寸的空隙，旨在尝试创造一项有趣的模块化设计，同时为人类创造最佳的未来生活环境。我们将这些结构块称之为 SMART BOX，它们可以在当前建筑物原始骨架结构的各个方向上活动，因此可以根据人们的需要实现任何方向的移动。

The big idea is stemmed from three main parts of a Tree. We made different sizes of voids in the structure to try to create an interesting and modular design and also the best future living environment for humans. The blocks, we call it the SMART BOX which can move in the various direction based on the original skeleton structure that existing building has, is able to move according to people's needs.

前文提到了建筑物内外的界限。在未来社会中，人与人之间的互动十分重要，因此所有开放空间均采取透明设计（智能交流集会）。正如树枝上生长着枝叶的树木一样，在未来，现有建筑物的结构块还将会继续增长。

The boundaries of the inside and outside of the building are mentioned, transparency is given to all the open spaces, the interaction between people is important in a future society (commune Smart Assembly). Just like a TREE with foliage growing over the branches, the blocks of the existing structure and will continue to grow in the future.

 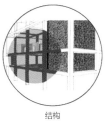 

绿墙
THE GREEN WALL

结构
THE STRUCTURE

公共空间 / 衔接空间
PUBLIC SPACE/CONNECTING SPACE

模块开发
MODULE DEVELOPMENT

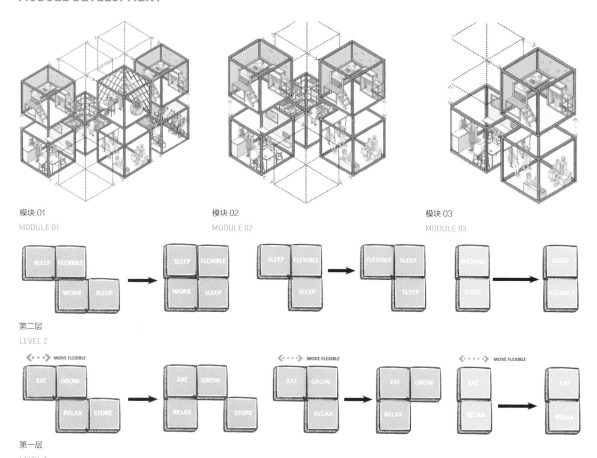

模块 01
MODULE 01

模块 02
MODULE 02

模块 03
MODULE 03

第二层
LEVEL 2

第一层
LEVEL 1

场景安排
ARRANGEMENT SCENARIOS

这些结构块是功能性空间，可以根据不同的需求在建筑物内移动。尤其是在公共空间内，功能将随着季节而变化，以满足每个季节的需要。
我们建议将已开发部分中的所有现有挡板取出，以便模块灵活移动；而不同的场景则体现了模块根据各个季节所做的相应安排。

The boxes are functional spaces which can move to adapt the changes of the demand within the building, especially within the public zone, the function will change throughout seasons to match with the need of each season.
We propose to take out all existing slabs for the developed part to give flexibility for the modules to smartly move. The scenarios elaborate the modules' arrangement according to each season.

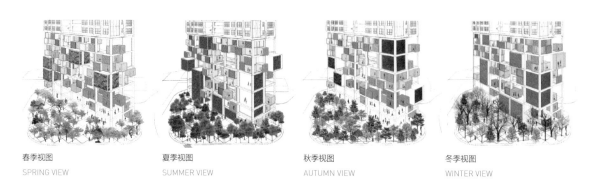

春季视图
SPRING VIEW

夏季视图
SUMMER VIEW

秋季视图
AUTUMN VIEW

冬季视图
WINTER VIEW

轴测平面
AXONOMETRIC PLAN

轴测剖面
AXONOMETRIC SECTION

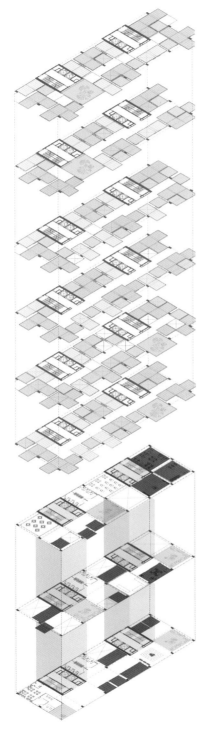

住宅　RESDENCE

原始结构	附加滑动结构
ORIGINAL STRUCTURE	ADDITIONAL SLIDING STRUCTURE
结构中的模块	可沿着结构移动的模块
MODULES WITHIN STRUCTURE	MOVEABLE MODULES ALONG STRUCTURE

立面视图
FACADE VIEW

住宅视图
RESIDENCE VIEW

鸟瞰图
BIRD VIEW

住宅视图
RESIDENCE VIEW

荣获奖项：入围奖　　　Awarded: Finalist Award

HUB——垂直智慧社区　　HUB —— Vertical Intelligent Community

团队成员：Nikolay Martynov　　　Team members: Nikolay Martynov

概念
CONCEPT

传统的住房供给模式与现代城市居民不断加快的生活节奏和日益增长的居住愿景产生了脱节。该中心是在数字时代背景下产生的创新共生概念。它是一种建立在共享经济，互联互通和协同合作原则上的新生活模式。
Traditional housing models are out of touch with constantly accelerating pace and aspirations of modern urban dwellers. HUB is an innovative co-living concept for the digital age. It is a new living model based on principles of shared economy, connectivity and collaboration.

该中心将办公、休闲和家庭生活融合在一个多样化，多层次的环境中。它是一个新的创造性的生活空间，能够促进居民之间的社会互动，培养新的创业理念，提供高品质的文化生活方式。
HUB merges office workplace, leisure and family life in a diverse multilayered environment. It is a new space for creative living that promotes social interaction among residents, fosters new entrepreneurial ideas and provides a high-quality cultural lifestyle.

工作 WORK

生活 LIVE

娱乐 ENTERTAINMENT

HUB VERTICAL INTELLIGENT COMMUNITY

智能城市生态系统
SMART URBAN ECOSYSTEM

该中心是空间、思想和资源共享以及在密集多样的城市环境中生活的物理和数字平台。该中心通过一个在线平台运营，该平台允许会员连接到创作者、企业家和居民的整个社区。
HUB is both physical and digital platform for sharing space, ideas, resources and living in dense and diverse urban environment. HUB is operated through an online platform which allows members to connect to the entire community of creators, entrepreneurs and residents.

该中心是一个基于预定的数字生态系统。通过中心在线门户预订微型单元、办公室和便利设施空间，该门户跟踪用户的时间和日程安排。租金率取决于空间的使用，并可根据使用强度增加或减少。

HUB is a subscription based digital ecosystem. Micro Units, Office and Amenity space are booked through HUB On-line portal, which tracks the user time and schedule. Rent rates depend on the use of space and can increase or decrease based on the intensity of use.

灵活的未来生活模式
FLEXIBLE MODEL FOR FUTURE LIVING

微型单元是未来生活的一种实验性住宅类型，便于在未来实现灵活的再安排。紧跟房地产市场的发展趋势 – 单元可以根据需要进行更换和更新。预制模块从工厂交付，并就地插入现有结构系统。居民可以使用不同尺寸的多个空间配置——从一个小工作室到一个宽敞的两居室公寓。

Micro Unit is an experimental residential typology for future living which provides flexibility for future re-arrangement. Following trends of real-estate market - Units can be replaced and updated on demand. Prefabricated modules are delivered from the factory and plugged in-place into the existing structural system. Several spatial configurations of different sizes are available to the residents - from a small studio to a spacious two-bedroom apartment.

模块化结构
MODULAR CONSTRUCTION

微型单元模块在工厂预制。将建成的单元运输到现场，安装到建筑物的现有结构系统中。现场施工时间可以随时间修改单元配置的最小值为限。模块化系统和工厂装配可以更好地控制质量，有效使用材料。

Micro Unit Modules are prefabricated at a factory. Fully built units are transported to the site and installed in place into existing structural system of the building. Time of construction on site is limited to the minimum which allows to modify the unit configuration over the time. Modular system and factory assembly allow better quality control and efficient use of materials.

内置家具
IN-BUILT FURNITURE

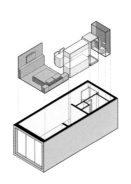

小尺寸的单元需要提高空间使用率。家具元件被开发成灵活内置模块，可以整合到微型单元模块中。在占用最小空间的同时，家具板块也创造了舒适的小型空间环境。

Units of small size require increased efficiency of the use of space. Furniture elements are developed as flexible built-in blocks which could be integrated into Micro Unit Modules. While occupying minimum space, furniture blocks create comfortable small-scale environment.

插入式 MEP 系统
PLUG-IN MEP SYSTEMS

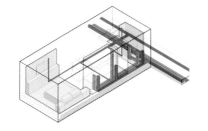

每个微型单元都配有内置的 MEP 系统。活动地板为管道、供暖和空调管道、电气和IT通信提供了空间。当在现场安装时，将微型单元的MEP系统插入中心脊柱，并连接到整个建筑的MEP网络。

Each Micro Unit is equipped with a built-in MEP system. Raised floor provides space for piping, heating and air conditioning ducts, electrical and IT communications. When installed at the site, Micro Unit's MEP systems will be plugged into the central spine and connected to the MEP network of overall building.

| 荣获奖项：入围奖 | Awarded: Finalist Award |

All You Need is BOX　　ALL YOU NEED IS BOX

团队成员：郑天宇，王谦　　Team members: Zheng Tianyu, Wang Qian
来自：马德里理工大学　　From: Universidad Politécnica de Madrid

概念
CONCEPT

抛弃定义！随形易变！
DEFINE NOTHING! FACILITATE EVERYTHING!

问题
如今的居住空间被功能严格定义，它被划分为起居室、卧室、厨房、厕所等等。这源于20世纪功能主义思想。然而，如果我们回溯历史，我们会发现居住空间排布的多样性。在古代，家是一个以火为中心的圆形小屋，室内并不存在额外的划分。"住宅是居住的机器。"依照这个准则，住宅被设计成一系列有具体功能的房间。比如法兰克福厨房就是一个这样想法的典范。在这个情况下，空间被划分为不同部分却不能被充分利用。这个问题将会变得异常严重，因为当代城市拥有巨大的人口密度。而且，空间也不能满足人们的需求。

Problem
Today's domestic space is always strictly defined by its functions as living room, bedroom, kitchen, restroom, etc. This is a heritage of the 20th century's functionalism. However, if we look back into the history, we can find the distribution of domestic space was surprisingly different. In the ancient time, a home is a round cottage with a fire in the center without any subdivision in the interior. "Une maison est une machine-à-habiter." Following this instruction, a home was designed as a series of rooms for specific uses. The Frankfurt Kitchen was an excellent practice of this idea. Under this practice, space will be divided into different parts and cannot be used sufficiently. This problem will be more serious because of the extreme high population density in contemporary cities. Secondly, space will no longer satisfy people's changing demands.

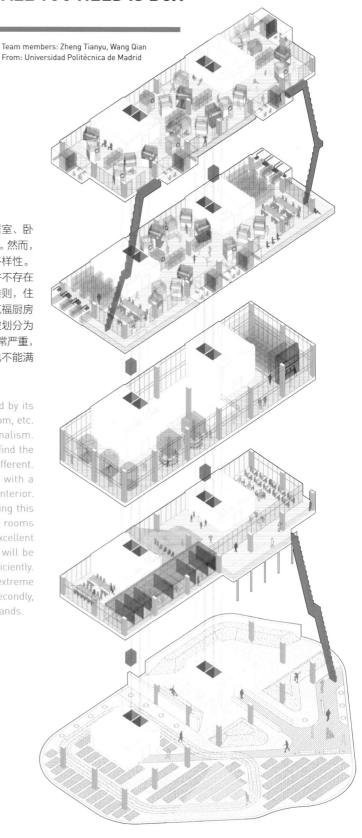

策略

因此，我们的方案将所有居住的服务空间浓缩在几种盒子之中，并悬挂于建筑外侧。由此获得了多用途无等级的室内空间。居住单元的盒子是可移动的，就如同其他家居一样，因此，所有可能的活动能够在这个空间发生。在整个建筑的公共部分，我们设计了不同用途的盒子，比如工作，学习，购物，等等。所有这些盒子都在工厂预制，之后在置入在建筑之中。他们可以自由的去除和替换。

Strategy

In our proposal, we condense all livable domestic space into several boxes and hang them on the outside. Therefore, we realize a multipurpose interior space with no hierarchy, or spatial definition. The residential boxes are movable, so are other furniture. Due to this, all possible activities can happen in the space. In the public area, we design a series of boxes for diverse uses like working, studying, shopping, etc. All these boxes are prefabricated in factories and then inserted into structures. They can be freely removed and replaced as required.

一无所有！分享一切！
OWN NOTHING! SHARE EVERYTHING!

问题

在《没有厨房的城市》中，Anna Puigjaner 研究了纽约 1871 到 1929 年间的出现的一种酒店式公寓的居住类型。这些公寓不含有私人厨房。相反，只在底层有公共厨房为所有居民提供食物。我们可以发现，公共的居住服务设施能够解放个人的居住空间。当代的信息技术和便捷的运输系统使得共享资源更加轻易获取，我们可以提出这样疑问：当所有食物能在 30 分钟内送达的时候，为什么我们还需要一个厨房？甚至：如果任何所需的东西都能被立即运送给我们，为什么我们还要在家放置一些使用率极低的物品？

Problem

In Kitchenless City, Anna Puigjaner studied the Hotel Apartment as a typology of housing existing in New York from 1871 to 1929. Apartments in these buildings didn't have a private kitchen. Instead, there was a collective kitchen in the basement offering foods for all residents. It is interesting to find that collective domestic services can liberate private living space. While contemporary information technology and the efficient delivery system make the collective resources easy to find and fast to get, it is safe to ask: why we need a kitchen when any food can be sent to your home in 30 minutes or less? Why do we settle things of little use in home if anything we need can be delivered to us once we need?

策略

在这个社区，居民就如同游牧民族一样居住在可移动的盒子中。我们尝试将个人空间浓缩到最小。每个居住单元只包括睡眠舱和个人物品储藏室。所有其他空间都是共享的：厕所、厨房、洗衣，等等。其他家用物品，比如吸尘器和熨斗都在一个公用橱柜内。所有人都可以通过 APP 来申请使用。

Strategy

In this community, residents live in a movable box like a nomad. We reduce the private space to its minimum. Every residential unit only contains a sleeping tank and a cloakroom to store personal effects. All other space is shared: the toilet, the kitchen, the laundry and so on. Other household items, like dust collector and iron, are collected in a cabinet. Residents can borrow them through APP to apply for occasional use.

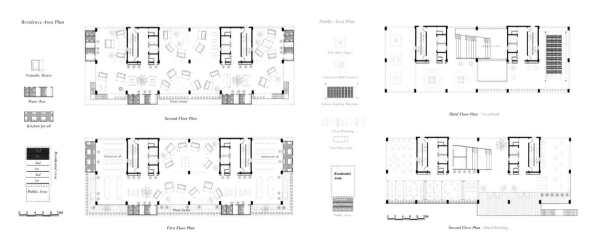

居住单元
THE LIVING UNIT

预制化
PREFABRICATED

所有居住单元以同样的方式在工厂预制，但是住户可以定制单元的具体尺寸和颜色。现提供两种可选尺寸：2m×2m×2.7m（单人房）和2m×2.5m×2.7m（双人房）。

All the units are fabricated in factories in a same mode. But the resident can customize the size and color of the unit. We now have two optional size: 2m×2m×2.7m for singles and 2m×2.5m×2.7m for couples.

游牧化
NOMADIC

所有居住单元均可动，可以以任意角度放置在任何地方。住户可以自由地聚集和撤离单元。因此，可以产生基于共同兴趣或者相似生活方式的微型社区。

All the living units are movable, and they can be placed anywhere in any angle. People can freely get together and withdraw these units. Hence, small communities can be formed for common interests or similar life pattern.

互动化
INTERACTIVE

单元被通电玻璃制成的屏幕包裹，人们可以在上面发布文字或图片。同时，这个表皮也定义了个人物品的储存空间。当屏幕模式关闭，表皮转换成玻璃时，它也作为个人展览空间。

The box is enveloped with a screen on which people can publish any words and photos. Meanwhile, the envelop also defines the storage space of personal things, which is an exhibition gallery when the envelope mode is switched from the screen to the glass.

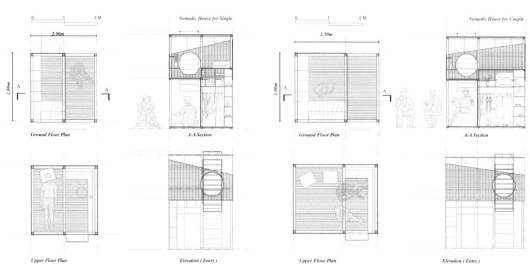

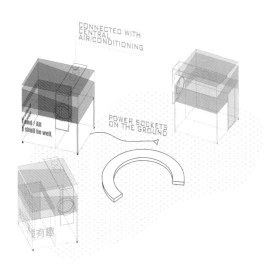
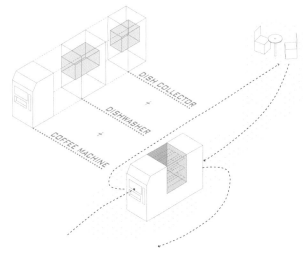

物联网系统设计：我的智能家居
INTERNET OF THINGS:MY SMART HOME

所谓物联网，是指将所有能独立行使使用功能的物品通过互联网联系起来，以此实现信息交换和资源调配。受这个思路的启发，我们为这个社区设计了一个智能生活管理系统。这个系统分为两部分：第一，网上预定系统。在住户入住之前，需要在网页上注册和登陆，并定制属于自己的居住单元。第二部分是一个整合了所有社区内公共服务的手机应用：My Smart Home。入住后，在手机上安装该应用并登陆，就可以管理自己的居住空间，以及线上预定和使用所有社区内的公共服务。在首页上点击对应的图标，就可以进入每个机器的服务页面，用户可以订餐、购物、租用办公空间等等。

The idea of IOT is to integrate the physical devices, whose functions can be used individually, into a computer-based system, thus realizing information exchange and resource allocation. Inspired by this idea, we intend to design Smart Life Management System. The system contains two parts: first is the online booking system. Before households move in, they need to register and log on the web page, and customize their own living units. The second part is a mobile application that integrates all public services in the community.: My Smart Home. After checking in, download the App in your phone and log on to manage your living space, book online as well as use public services. By clicking the corresponding icon on the home page to enter the service page of each machine, users can order meals, shop, rent office space and so on.

荣获奖项：入围奖　　Awarded: Finalist Award

定制生活，从盖房子开始…… Customized Life Starts with Building...

团队成员：周茵，唐松，黄帅　　Team members: Zhou Yin, Tang Song, Huang Shuai
来自：IN-OUT Studio　　From: IN-OUT Studio

未来生活
FUTURE LIFE

完整生活
COMPLETE LIFE

通过互联网未来人们在家里或者社区就可以完成办公、医疗、健身、购物，甚至去动物园等各种行为。
In the future, people will be able to carry out all kinds of activities, such as office work, medical treatment, fitness, shopping, even going to the zoo, at home or in the community.

定制生活
CUSTOMIZING LIFE

去产品超市选定"你的"住宅，甚至可以以旧换新。在未来，只有定制到人的住宅才能最大限度满足人们个性化和多样性的需求。
Go to the product supermarket to choose your "house", or even replace old ones for new ones. In the future, only customized housing can satisfy the needs of people's individuality and diversity.

和植物一起生活
LIVING WITH PLANTS

从"家"单元开始"编织"绿色，再到社区再到街道和城市，逐步放大的密集绿色网络将植物与生活编织在一起。
We start from the "home" unit to "weave" green, then to the community and then to the streets and cities, gradually enlarging the dense green network to weave plants and life together.

来居住概念——定制住宅
THE CONCEPT OF FUTURE RESIDENCE—CUSTOMIZED HOUSING

1. 用大规模定制解决高效率和个性化需求的矛盾
高效率需要标准化、统一化和重复使用，个性化需求强调个体，要与众不同。两者的矛盾可以用大规模定制来解决。集合、分解，让效率归效率需求归需求。尤其适用于大量需求集中的高密度中心城市。

1. Use mass customization to solve the contradiction between high efficiency and individualized demand.
Efficiency requires standardization, unification and reuse, while personalized needs emphasize individuality and difference. The contradiction between them can be solved by mass customization. Let the efficiency be efficiency and demand be demand by methods of set and disassembly. It is especially suitable for high density central cities with large demand concentration.

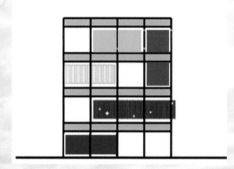 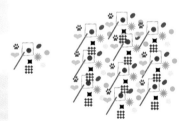 

2. 定制住宅整体系统
流程：开发商拿地，完成满足用地条件的规划布局。建筑部分只需完成标准框架和交通筒。与此同时开始收集客户需求，需求收集完毕后，由产品设计师利用大数据进行编排整理。最终客户的需求在各种数量不同组合的"盒子"里实现。

2. The Whole System of Custom Housing
Process: The developers take the land to complete the planning and layout of land conditions. The construction part only needs to complete the standard framework and traffic container. At the same time, customer requirements are collected, and after the requirements are collected, the product designer arranges the requirements with large data. The ultimate customer needs are implemented in various boxes of different combinations.

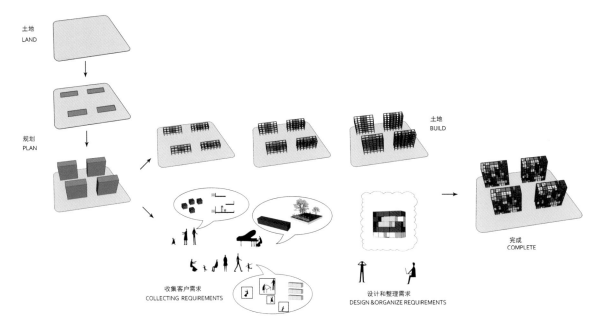

3. 未来居住单元

1）理念：标准框架 + 定制盒子 + 绿隙。

2）虚拟客户

A 先生："我想购一套婚房，和未来的太太两个人居住。需要 3 个盒子，卧室在二层；我们希望有大露台，可以种树种花。"

B 女士："我和先生都是音乐家，我们有两个孩子。房子需要 6 个盒子；一层有大的工作空间，卧室安排在楼上；我们还希望拥有大面积的露台。"

C 先生："我们一家六口人，除了我们夫妻俩还有两位老人和一对子女。我会购买 8 个盒子，布局上希望三代人都能有各自的生活空间，并且还需要一个大面积的共享空间。"

3）标准盒子单元：4.2米×4.2米×3.3米（高）。

3.FUTURE RESIDENTIAL UNIT

1) Concept: Standard Frame + Customized Box + Green Space

2) Virtual customer

Mr.A: "I want a house to live with my future wife. We need 3 boxes. The bedroom is on the second floor, and we want to have a big balcony, where we can plant trees and flowers."

Mrs. B: "my husband and I are musicians. We have two children. We needs 6 boxes; a large working space on the first floor with bedrooms upstairs; and we want a large terrace.

Mr. C: "there are six people in our family. Besides our husband and wife, there are two old people and a couple of children. I'm going to buy 8 boxes, and the layout should be planned allow three generations to have their own living space and a large area of shared space.

3) Standard Custom Box: 4.2m×4.2m×3.3m (Height)

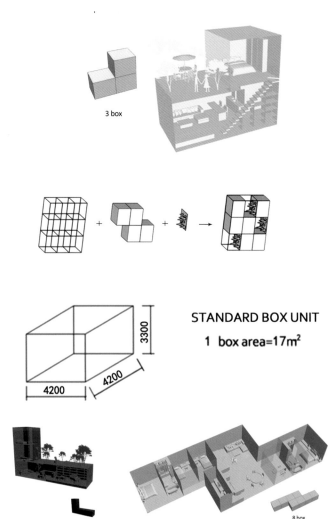

4. 像树一样生长
4.GROWING LIKE A TREE

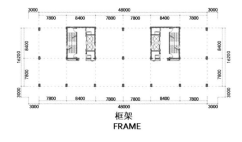

框架
FRAME

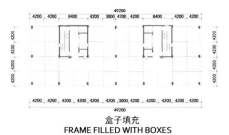

盒子填充
FRAME FILLED WITH BOXES

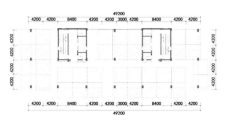

盒子填充
FRAME FILLED WITH BOXES

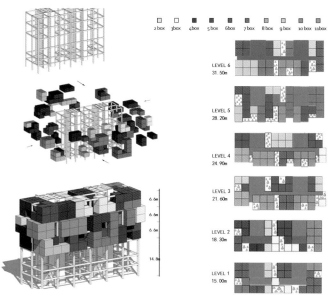

5. 标准盒子构造说明

盒子的结构和内外装材料构造都最大限度地采用标准化预制装配式系统，以提高效率，节约能源，并且便于维修和更换。用户甚至可以随时更换盒子。内部的相关技术有双层天棚，可用于埋设管线；架空楼板，实现同层排水及负压式新风系统；预制干式内墙系统，实现内部空间灵活变化。

5. CONSTRUCTION SPECIFICATION OF STANDARD BOX

Standardized prefabricated assembling systems are maximally used in the box's structure and internal and external materials to improve the efficiency, energy conservation, and ease of maintenance and replacement. Users can even change boxes at any time. Internal related technologies involve double ceiling, which can be used to bury pipelines; overhead floor, to achieve the same floor drainage and negative pressure fresh air system; prefabricated dry wall system, to achieve flexible changes in internal space.

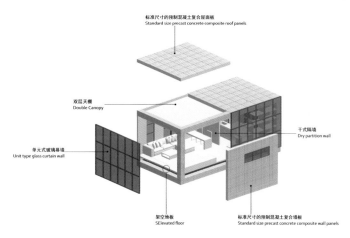

荣获奖项：入围奖　　　　　　　　Awarded: Finalist Award

"拼图式"垂直花园　　　Jigsaw Vertical Garden

团队成员：骆肇阳　　　　　　　　Team members: Luo Zhaoyang
来自：麻省理工学院　　　　　　　From: Massachusetts Institute of Technology

"拼图式"垂直系统 + 人工智能算法
JIGSAW VERTICAL SYSTEM + AI ALGORITHM

拼图式系统，灵感来源于七巧板，七巧板可以无缝拼接并形成不同形状。本方案以七巧板拼图为原型，设计出不同的单元模块，将这些模块赋予功能，譬如居住单元模块有起居室、厨房、卧室、卫生间等，公共空间模块有超市、餐厅、书店、创客空间等，并制定不同的面积空间模块，以满足不同年龄段、经济水平的人群居住计划需求，功能模块根据居住者使用需求进行设立与更新，能够可持续使用；并将这些模块垂直设立成树状系统，以最大化利用空间，满足高密度需求。结合人工智能算法，利用人工神经网络模型进行方案预设学习，先利用遗传算法进行形态最优解的提取，以使用者的相关信息譬如居住人数、居住年龄段、经济能力等作为设计参量（输入值），求取最为理想的居住组合模式并将其转译为设计结果（输出值），接着以"输入——输出值"作为映射，对人工神经网络模型进行训练，以及数据测试，最终获取更为迅速的方案生成引擎，且在物联网的帮助下，数据越庞大，该引擎运算越快。

The jigsaw system is inspired by the jigsaw puzzle, which can be seamlessly stitched together to form different shapes. This program uses the jigsaw puzzle as a prototype to design different unit modules, and these modules are given functions. For example, the living unit module has living room, kitchen, bedroom, bathroom, etc. The public space module has supermarket, restaurant, bookstore, maker space, etc. And in order to develop different area and space modules to meet the needs of people of different ages and economic levels, the functional modules are set up and updated according to the needs of the occupants, and can be used sustainably; and these modules are vertically set up into a tree system, to maximize the use of space to meet high-density needs. Combined with artificial intelligence algorithm, the artificial neural network model is used to predict the program. Firstly, the genetic algorithm is used to extract the optimal shape solution. The user's relevant information, such as the number of residents, the age of residents, economic background, etc., are used as design parameters. (Value), to find the most ideal residential combination mode and translate it into design results (output value), and then use the "input-output value" as a mapping for the artificial neural network model training, and data testing, and finally obtain a Faster Scheme Generation Engine. With the help of the Internet of Things, the larger the data is, the faster the engine operates.

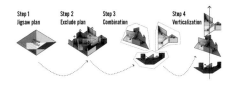
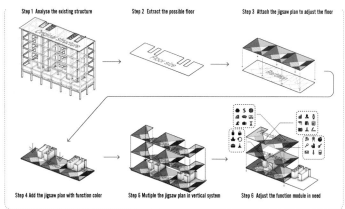

公共空间的模块组合
PUBLIC PLACE MODULE

公共空间设计采用嵌入式设计，设计一系列嵌入式公共单元，包括小型研发中心、创客空间、商务会议、图书馆、展厅、超市、餐厅、咖啡厅、茶室、健身、电影院、花园等功能，并能实现最大程度的灵活可变性。

The public space design uses embedded design to design a series of embedded public units, including small R&D centers, maker space, business meeting, libraries, showrooms, supermarkets, resturants, cafes, tea rooms, fitness, cinemas, gardens, etc. And to achieve maxmum flexibility.

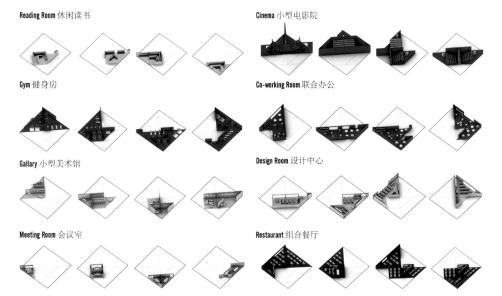

李兴钢
Li Xinggang

全国工程勘察设计大师，中国建筑设计研究院有限公司总建筑师
李兴钢建筑工作室主持人

the National Master of Engineering Survey and Design, the Chief Architect of China Architecture Design & Research Group, the director of Atelier Li Xinggang

垂直社区的生态定制
Ecological Customization of Vertical Community

随着城市发展，人口不断聚集增加，人均土地资源不断减少，当代的人们不再有条件生活在自然山水之间，特别是近年来，高度城市化导致交通问题、空气质量问题及人居生活空间等问题的恶化。一方面是城市无休止推土机式地扩展，不顾自然环境的保持和与城市空间的协调，形成"千城一面"；另一方面，原本的历史城区又处于一种不合理的超密集状态，人与自然越来越隔离疏远，虽然表面上经济生活水准提高了，但人们实际上已经不再拥有理想的生活环境。

Along with the development of cities, the population has kept gathering and increasing while the per capita land resource has kept decreasing; contemporary people, therefore, have no conditions to live between natural landscapes any more. Especially, in recent years, high-degree urbanization has induced the deterioration of the problems with traffic, air quality, human settlement and living space, etc. On the one hand, the cities have kept expanding endlessly like "being driven by bulldozers", regardless of the maintenance of the natural environment and the coordination of urban spaces, forming "thousands of cities with the same pattern"; on the other hand, original historical urban areas are in an unreasonable super-dense state, isolating people further from the nature and depriving them from an ideal living environment, despite the superficially improved economic standard of living.

这次竞赛聚焦的话题正是高密度城市中的垂直社区。在这类垂直社区中，公共生活、社区生活和居住生活相混合，绿色生态也必须向垂直方向立体延伸。此种现实下，建筑师怎样为人们营造出自然惬意的理想空间，满足人性对自然和生态的追求？参赛者做出了积极努力的尝试，在设计中大胆想象，把可变设计、智慧管理、生态居住的定制化等科技手段融入建筑之中，为城市建筑的未来提供了思路。

The focal topic of this contest is right the vertical communities in high-density cities. Green ecology must extend vertically in a three-dimensional way in such vertical communities where public life, community life and residential life are mixed mutually. Under such practical condition, how could architects create a natural, comfortable and ideal space for people and meet the human being's pursuit of the nature and ecology? The participants have made positive and arduous attempts, and imagined boldly during the design contest, and integrated the technological means such as variable design, intelligent management, and ecological residence customization, etc. into the buildings, providing a reference for the future of urban buildings.

人们居住的基本需求，一方面包括家庭内部活动，比如用餐、睡眠、娱乐、交流等等，另一方面，人也需要与自然接触，即便在高层居住综合体中，人们也向往通过某种模拟的自然生态环境获得精神上的享受，在日常生活中体验从喧嚣到静谧的"自然"回归。当下，各种信息技术、数据技术和人工智能技术，使得科技更有可能帮助人们满足理想化的需求，甚至使下一代建筑重构新的生活环境，也因此可能改善甚至解决影响人们生活质量的各种问题，如空气污染、高密度居住、远离自然、邻里关系、劳逸时间分配等等。

People's basic demand on residence includes, on the one hand, the internal activities of families, such as dining, sleep, entertainment, and communication, etc., on the other hand, people need contact with the nature. Even living in high-rise residential complexes, people also aspire to obtain the spiritual enjoyment through some simulation of natural ecological environment, and to experience the return from daily noisy life to silent "nature". Nowadays, various information technologies, data technologies, and AI technologies will possibly help to satisfy people's ideal demands and even allow the next-generation buildings to create a new living environment. Consequently, this will more possibly improve and even solve various problems affecting the quality of people's life, such as air pollution, high-density residence, isolation from the nature, neighborhood, allocation of labor and leisure time, etc.

作为建筑师，应当更敏锐地察觉到新时代人们生活中的各种问题，同时了解新技术应用的可能性，并将这两者更有效地结合，使技术更好地服务于建筑。在未来，建筑师的工作应该是一体化的，他需要了解或掌握各方面的专业知识，不能因为技术上的障碍影响对于建筑整体性的思考。科技将成为未来建筑整体性的一部分，互联网数据、智能交互、智慧管理，生态系统等等都将是下一代建筑最的显著趋势。

As architects, we should be more acutely aware of various problems in people's life in the new era, and meanwhile, understand the possibility of the application of new technologies, combine them more effectively, and thus make the technologies better serve buildings. In the future, the work of an architect should be integrated. The architect should know or grasp the professional knowledge in each relevant aspect, and should not be affected when thinking about building integrity due to technical barriers. Technologies will become a part of the integrity of future buildings: Internet data, intelligent interaction, intelligent management, and ecological system, etc. will all be the most remarkable trends of the next-generation buildings.

此次竞赛关注下一代建筑在垂直社区中的智慧创新，不仅与中国的当代现实息息相关，即使在世界范围内也体现出领先的理念与追求。我建议让竞赛的结果真正落地实施，把获一等奖的作品作为主要参考主体，并结合其他获奖作品的某些创意，通过与设计团队的合作深化，使之成为一个被实现的设计，这样才能真正实现这次竞赛独特的意义和价值。未来，希望可以看到更多大胆、富有想象的提案，并能通过更多的竞赛，让建筑设计更加符合社会与人的发，在当代条件下实现理想生活空间。

This contest is focused on the intelligent innovation of next-generation buildings in vertical communities. It is not only closely related to the contemporary reality of China, but also embodies the leading concept and pursuit in the global scope. I suggest truly implementing the results of this contest: take the works of the first prize as the major references and turn them to reality in combination with some originalities of other awarded works through the deepened cooperation with the design team, so as to truly realize the unique significance and value of this contest. In the future, we hope to see more bold and imaginative proposals, and through more contests, we hope that building designs will better meet the development of the society and human beings and create an ideal living space under the contemporary conditions.

张应鹏
Zhang Yingpeng

江苏省设计大师,九城都市建筑设计有限公司总建筑师,东南大学、浙江大学、华中科技大学兼职教授

Design Master of Jiangsu Province. Chief Architect of 9 Town Studio, Adjunct Professor of Southeast Univ., Zhejiang Univ. and Huazhong Univ. of Science and Technology

高适应性与功能化的未来建筑
Architecture in the Future Would be Highly Adaptable and Have Various Functions

当代建筑面临智能化和大数据等科技的快速发展,更重要的是,科技的发展与建筑发展,两者越来越紧密相随,以致同步更新。由此,建筑迎来了全新的变化。

Contemporary architecture develops against a backdrop of fast advancing technologies including intellectualization and big data. More importantly, the development of technologies and that of architecture are increasingly correlated and even update synchronously. Thus the architecture industry will see brand new changes.

传统定义的建筑,本身是一个物理空间,从一开始就负责解决两个问题:一是帮助提供改善生存的物质条件,二是用来满足精神需求,因为建筑会逐渐聚合成城市形态,组织人们的生活方式和交往方式。首先,作为物质化的物理空间,科技的发展会给建筑发展更有力的支撑,建筑材料和建筑手段与科技交融将越来越紧密。同时,在组织人居生活方式等精神需求方面,下一代建筑必将是更智慧的建筑。人工智能手段直接影响到建筑,进而影响人们的生活与城市的组织。

According to traditional definition, architecture is a physical space. It should meet two needs at the first place — improving physical conditions and satisfying spiritual needs. Buildings would gradually form the environment of a city and thus influence peoples' ways of living and interacting. Firstly, technological development would provide more powerful support for buildings as the physical space. Building materials and methods would integrate more closely with architecture of the next generation. Secondly, in terms of living style and other spiritual needs; architecture of the future is surely become more intelligent. Artificial intelligence would directly influence architecture and thus affect people's lives and city structures.

对传统建筑目标的追求,从古希腊时期开始,发展到现在已经非常成熟且受到公认。古往今来,人们衡量建筑的标准是实用、坚固与美观。未来,这三条标准将被颠覆与代替:与实用相比,是高度的适应性;与坚固相比,是主动出击的安全性;与美观相比,是可变的个性化。这三点,无疑都与当代科技信息科技与智能化的发展息息相关,建筑将成为一种全新的智慧体。

Pursuit for traditional architecture started from as early as the ancient Greek times. After years of development, the traditional architectural concept is very mature and well accepted by people. For years, the three standards for buildings are commodity, firmness, and delight. In the future, these three standards will be replaced by high adaptability, proactive security and flexible personalization respectively. These three aspects are closely related to the development of scientific, information and intelligent technologies in this era. In the future, construction would become a brand new form of intelligent being.

首先，理解传统建筑的功能，目前还局限在具体静态的物理空间上，也即满足生活中一些固定的实用功能。颠覆这种传统实用性的，是现代的适应性。通过当前先进的计算机模拟、数据采集交互和智能化技术，原本固定静态的物理空间可以根据居住者的需求和生活方式的改变进行变化，从而具备了更强的弹性与适应性。

First of all, our understanding on traditional buildings is still limited on specific and static physical space with functions that meet certain practical needs in people's lives. This traditional perception will be replaced by the requirement for modern adaptability. With the aid of advanced computational modeling, data collection, interaction and intelligent technologies; originally static physical space can also alter in response to changes in the needs and lifestyles of residents. Thus it will become more flexible and adaptive.

其次，过去的建筑必须依靠提高强度来达到坚固的目的。然而，强度是有限的，即便希腊建筑罗马建筑用石材把墙做得很厚，也无法面对地震、火灾等一些不可抗力。过去的建筑使用的是肌肉，未来的建筑却需要依靠智慧。对于抗震，可以设计阻离抗震，把地震力消减掉。对于排烟，用自动喷淋、自动报警实现消防目的。正如如今汽车上的安全气囊、自动平衡系统、无人驾驶系统运用了智慧的新技术来避免碰撞，未来的建筑也需要高端的智慧系统，来达到主动性的安全防御效果。

Second of all, people try to enhance firmness by increasing intensity of contraction, yet intensity is limited. No matter how thick the stone walls of Greek and Roman buildings were, they could not weather forceful majeure events like earthquakes and fire hazards. Architecture in the past relied on labor, yet buildings of the future depend on intelligence. We could design quake-proof buildings with dampers to offset the power of the earthquake. Moreover, we could use automatic sprinklers to eject smoke and install automatic alert systems for fire protection. Just like automobiles today, which are equipped with air bags, auto balance systems and autonomous driving systems among other new intelligent technologies to avoid collision. Buildings in the future also need highly intelligent systems for automatic security mechanisms.

最后，在传统的美学观念里，美是绝对的。美是有终极定义的概念，文艺复兴时期，甚至可以用数学计算出美的形式。但是随着时代的发展及多样化的需求诞生，个性化越来越替代了以往统一绝对的美。如何实现这一个性化？一方面是通过不同的建筑师的创造，根据建筑师的理解和审美，不断发现独特个性的东西。另一方面，则要通过现代科技带来的新的技术支撑与运用，让建筑具有更多可以实现个性化的手段。

At last, the definition of beauty is fixed in the traditional concept of aesthetics. There is ultimately a fixed definition for beauty and it could even be calculated by a mathematical method during the Renaissance Period. Yet with the development of times and the emergence of diversified needs, the definition of beauty would become more personalized rather than unified and fixed. How to meet the need for personalization? On the one hand, architects need to make innovations and keep seeking for distinctive elements according to their own understanding and taste. On the other hand, we need modern technological support and applications to add more customizable features to buildings.

建筑的标准将随着时代而变化，这一变化的终极目的是服务于人，把人不断地变化的需求设计到建筑中，与人进行互动、变得更多元、更具适应性。虽然未来并不可知，但趋势已然存在。作为建筑师，我一直在期待，期待下一代建筑的颠覆性变革。

The standards for buildings may change in different eras, yet the ultimate purpose remains the same, namely, serving people's needs. We need to factor in people's changing demands into architectural design and enable buildings to interact with people and become more diversified and adaptive. Although the future is not predictable, the trend already existed. As an architect, I am always expecting subversive revolutions of architecture in the next generation.

Michael U. Hensel

奥斯陆建筑与设计学院 RCAT 实验室主任，OCEAN 设计研究协会的创始人及董事会成员

Director of RCAT - Research Center for Architecture and Tectonics of Oslo school of Architecture and Design, Co-founder of OCEAN Design Research Association

超越隐喻，思考未来
Go Beyond the Metaphor When Reflecting on the Future

越来越多的建筑以快速增长的节奏爬满了我们的地表，以致影响到它所在的生物与物理环境，甚至开始破坏物种和生态系统。

At a fast increasing pace, more and more buildings and constructions take up the surface area on our planet and in result affect the local bio-physical environment, and even begin to destroy local species and ecosystems.

此次关于下一代建筑的竞赛名为"智慧树"。关于"树"，我们思考的是：把树当成一种隐喻、类比、相似物、关联物，还或仅是其字面含义？在我看来，当理解超出"树之隐喻"的范畴时，才会有更多收获。我们或许会从树身上学到一些东西，以此作为使建筑适应当地环境和生态的契机。

The theme of this "Next Architecture Award" is "Smart Tree". With regard to "trees", the question is whether trees are referred to as a metaphor, analogy, analogue, or in a literal way? In my view, more is to be gained when going beyond the scope of a metaphor. Trees may provide us with an opportunity to learn how to adapt to the local environment and ecology.

在这个背景下，我们是否考虑过人类发明和建造的建筑也可以为树木的生长提供环境？或打算思考一下建筑的景观规划生产所需的植物和树木？或者，是否有必要连接因景观隔离带而产生的绿色走廊，以及树木可能在其中发挥的作用？

In this context, have we considered that the constructions invented and erected by humans can also provide environments for trees to grow and what the many positive contributions of urban trees and forests are? Or is there the necessity to connect the green corridors created by protected landscape zones, as well as what's the role trees might play in them?

实际上，这些只是在"树之隐喻"中，可能出现的几个问题。除了树，我们是否思考过生态系统所需的各种事物？
These are but a few of the questions and potentials that arise when going beyond the metaphor. In addition to trees, have we considered the various conditions that ecosystem require?

然后，在具体实施的方法层面，我们还需考虑用什么来超越"树之隐喻"，从而为设计注入更多的智慧。我们如何将有关环境的数据转化为能最终用于设计的有效信息与知识？这需要多种模式下的数据收集、信息构建、以及多标量、多领域的建模过程。
Furthermore, at the level of the implementation method, we also need to consider what is required from a methodological perspective to transcend the tree as metaphor and to bring more intelligence to the design. How can we translate environmentally relevant data into useful information, knowledge and ultimately utility for design. This requires multi-modal data collection, information structuring, and multi-scalar and multi-domain modeling processes.

对"生物"而言，信息具有比通常所知更重要的作用。然而，我们到底需要什么样的输入数据，用于以信息为基础的设计？何种类型的高级可视化技术，能够验证和开发基于信息的设计呢？
In biology information plays a much greater role than is usually acknowledged. Yet, what kind of data do we require for information-based design? And what types of advanced visualization technologies can serve to validate and develop information-based design?

同样，我们还需要思考：如何向建筑师、其他利益相关者以及建筑物的使用者提供必要的信息，以便将可持续发展性延续到环境的长期关注和维护中？该如何处理影响设计且反过来受设计影响的相互作用的复杂因素？在迭代建模过程中，我们可以用不同的方式来关联建筑部件吗？在这些过程中，设计和分析之间的迭代逻辑是如何设定的？
Similarly, we need to think about how do we provide necessary information to the architect, other stakeholders and users of the buildings we design such that sustainability can be extended to long-term care and maintenance of the environment. How can we cope with a greater complexity of correlating and interacting factors that influence the design and are in turn influenced by it? Can we correlate buildings parts in different ways in iterative modeling processes? How is the iterative logic between design and analysis set in these processes?

从景观设计的视角来看待大型项目，需要思考的是：建筑应如何与地形和微气候的变化相结合？哪种迭代过程能以关联的方式来服务内部和外部气候？当需要修建大量的建筑物时，是否即使在不利地形条件下，也能保持一定比例的生态网络，且不改变当地的生态、土壤和水的状况？哪种迭代过程可以有助于将跨尺度的分析周期和生成周期联系起来？如何将多个分析链，连接到项目的形态生成上，从而既涉及该给定项目又涉及当地环境、气候和生态系统的不同方面？有可能在建造中摒弃推土机和炸药吗？如何停止以非可持续的方式进行的对土地大量而有害的消耗行为？
When looking at large projects from a landscape design perspective, one needs to think about how can architecture be combined and embedded with existing topography and micro-climate. And when a large number of buildings need to be provided, can this take place on adverse terrain conditions while maintaining local ecological, soil and water conditions? Which kind of iterative processes might help to link analytical and generative cycles across scales? How can we link multiple analyses chains into the form-generation of a given project that pertain both the object as well as different aspects of the local environment, climate and ecosystems? Is it possible to leave the bulldozers and the dynamite out of the game? How to stop the massive and detrimental consumption of land in an unsustainable manner?

这一系列的问题，正是"智慧树竞赛"可能思考的，也是下一代建筑应该思考的。
These series of questions may be at the center of what the "Smart Tree" competition" may ask for when looking beyond metaphors, and constitutes in may ways what the next generation of architecture should think about.

Chapter of Smart Community Life

智慧社区生活篇

荣获奖项：二等奖　　　　Awarded: Second Prize

"DNA"双螺旋垂直街巷 ——街区重构
"DNA" Double Helix Vertical Street-Block Reconstruction

团队成员：刘志军、徐义飞、袁雷、孙韵雯
来自：江苏省建筑设计研究院有限公司

Team members: Liu Zhijun, Xu Yifei, Yuan Lei, Sun Yunwen
From: Jiangsu Provincial Architectural D&R Institute LTD

"垂直智慧社区"，致力于解决当下存在的高密度城市下存在的问题，既是"互联网+"、新型材料与能源开发以及建筑功能需求的不断提升下的探索，也是对可预见未来生活的理解与回应。公共性、流动性、可变性、智慧性是其核心特征点，也是"下一代建筑"的本质性格。

The "Vertical Smart Community" is dedicated to solving the problems existing in the high-density cities that exist today. It is not only the exploration of "Internet +", new materials and energy development, and the increasing demand for architectural functions, but also an understanding and response of the foreseeable future life. Publicity, mobility, variability, and intelligence are the core characteristics of the society, and the essential character of the "Next Architecture."

场景景观的处理与建筑的两条路径相关联，一层形成森林、绿地，营地通过路径从森林上空自然的回家，改变原有的明显层级关系。

The treatment of the site landscape is related to the two paths of the building. The first forms forests and green spaces, creating a home from the forest through the path, changing the original obvious hierarchical relationship.

设计基础
DESIGN BASIS

本案设计以当下街区模式、居住模式的"冰冷"感情、交往的闭塞为出发点，通过对传统街区模式的研究与学习，提出以"DNA"这种基因遗传与表达的方式，延续传统街区的"活力""交往空间"来重构当下街区模式，强调空间"交往、共享"的必要性并解决集约性土地利用的号召。

The design of the case is based on the current block mode, the "cold" feelings of the residential mode, and the occlusion of interaction. Through the research and study of the traditional block mode, we decide to come up with the idea of "DNA" gene inheritance and expression. On the one hand, reconstruction of the current block need to cling to the "vigor" and "communication space" of the traditional one; on the other hand, the current block mode emphasizes the need for "communication and sharing" of shape and addresses the call for intensive land use.

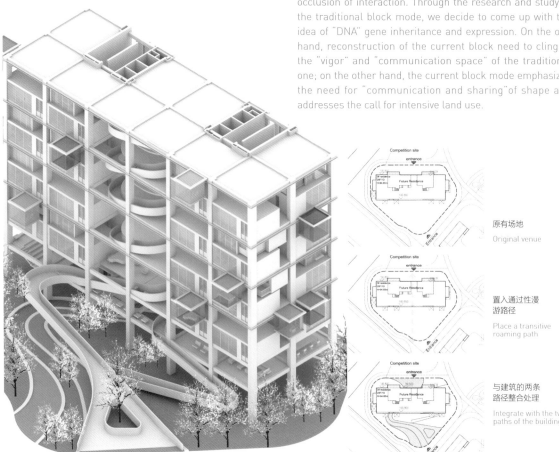

原有场地
Original venue

置入通过性漫游路径
Place a transitive roaming path

与建筑的两条路径整合处理
Integrate with the two paths of the building

设计概念
DESIGN CONCEPT

设计思路上对传统街区以及现在街区模式进行分析与比对，其中传统街区主要模式"街——家"，街道为事件的发生容器，具有很强的公共性与生活性；现代街区模式"街——小区——家"，在增强私密性的同时公共性也在降低，街道的生活性大大降低，甚至沦为单一的行走路径；设计试图以"垂直街巷"为概念来"取消""小区"环境重塑传统街区交往、氛围与活力并辅以当代技术与材料实现"公共性、流动性、可变性、智慧性"的垂直社区。

The design idea compares and analyses the traditional block and the current block mode. The traditional block mode "street-home"and the modern block mode "street-community-home" are designed to try "vertical street". The concept of "cancelling" the "community" environment reshapes the traditional neighborhoods, atmosphere and vitality and complements the vertical community of "publicity, mobility, variability and intelligence" with contemporary technology and materials.

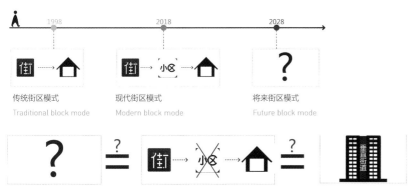

打破现有街区模式，重组回复传统街区模式（回复传统街区的交往空间与活力）
Breaking the existing block mode, reorganizing and restoring the traditional block mode(restoring the interaction sapce and vitality of traditional blocks)

设计细节
DESIGN DETAIL

分析传统街区中所发生的事件以及包含的功能，提取功能进行垂直化处理。
Analyze events that occur in traditional blocks and the functions they contain, and extract functions for vertical processing.

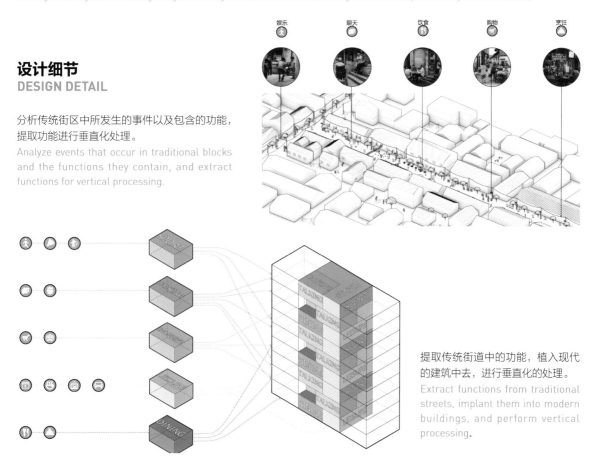

提取传统街道中的功能，植入现代的建筑中去，进行垂直化的处理。
Extract functions from traditional streets, implant them into modern buildings, and perform vertical processing.

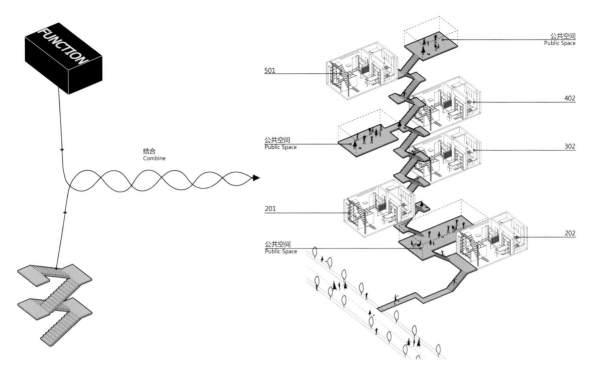

将传统街道中的功能与楼梯相结合，打破原来楼梯单一的垂直交通功能，使其变为承载公共活动的发生器，增强社会性，形成垂直社区
Combining the functions of traditional streets with stairs, breaking the single vertical traffic function of the original stairs, making it a generator carrying public activities, enhancing sociality and forming a vertical community

以"DNA"状的双螺旋街区为核心，公共功能置于两道街区之内，到达公共部分，定会发生"相遇"产生交流。
Taking the "DNA" double helix block as the core, the public function is placed within two blocks, and when it reaches the public part, "meeting" will occur to generate communication.

将原先的公共部分需要发生交往关系的功能与街区进行重组，并竖向插入建筑内部，每个单元与个体定会与街区发生直接关系。
Restructuring the function of the communicative relationship between the original public part and the block, and insert it vertically into the building. Each unit and individual will have a direct relationship with the block.

1. 设计范围
1.Design range

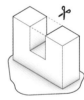

2. 将建筑中间部分打开，作为公共空间
2.Open the middle part of the builing as a public space

3. 在公共区域，设置一些作为活动的平台
3.In the public area,set up some platforms for activities

4. 运用楼梯将所有的平台连接，形成具有公共性的路径
4.Use stairs to connect all the platforms to form a common path

5. 提取传统山水画的路径
5.Extract the path in traditional landscape painting

6. 利用双螺旋组织公共活动空间，两条不交织的路径，促使人们在公共空间相遇，增加空间中人的交流
6.Take advantage of double helix to organize public activity space. Two non-interlaced paths encourage people to meet in pulic sapce and increase communication among people in the space

中央"垂直街巷"以"不定层高"进行处理，可根据使用者工作生活作息进行上下空间调节，在空间上形成联动以及出现"空间+"的理想状态，同时由于是双螺旋街道，平台在移动过程中始终是与街道相连接。

The central "vertical street" is treated with "indefinite layer height", which can adjust the upper and lower space according to the user's work and life, forming a linkage in space and an ideal state of "space+". As it is a double helix street, the platform is always connected with the street in the process of moving.

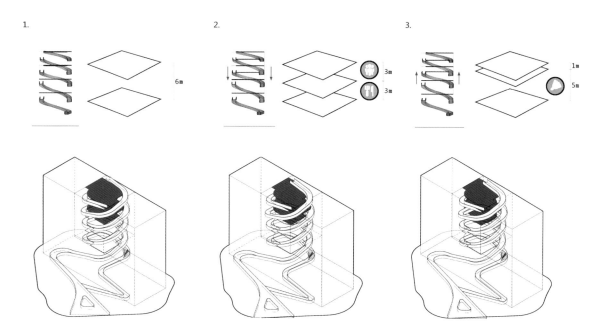

通过楼板的移动，可以满足不同时间段人们对公共空间的需求，来达到高密度城市下，公共空间的最大化利用。
Through the movement of the slab, people can meet the needs of public space in different time periods to achieve the maxmum use of public space in high-desity cities

基于集约、可变对单元体进行模块化研究，提出可"折叠式"的单元体模块，依据工作与生活作息以及在"互联网+"的条件下实现"居住"空间与"公共"空间的转化，与此同时整个建筑的里面形态也处于一个不断变换的状态，不同的状态又可以反映着建筑不同的状态信息，改变原有建筑单一的立面形态，使得建筑立面可变且成为信息表达体。

Based on the intensive and variable modularization of the unit body, a unit module that can be "folded" is proposed, which realizes the transformation of "living" space and "public" space according to work and life and under the condition of "Internet +". At the same time, the interior form of the building is also in a state of constant transformation, in which different states reflect different buildings. Such a design changes the single façade form of the original building, making the façade variable and expressive.

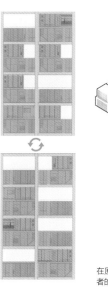

图例
居住空间 Living Space
公共空间 public space

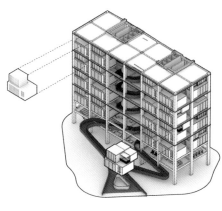

在原有的建筑框架中植入可变的住在单元，单元变化根据使用者的需求进行调整，实现居住空间与公共空间的转换。
Implanting variable residential units in the original building frame, unit changes are based on user needs to realize the conversion of living space and piblic sapce.

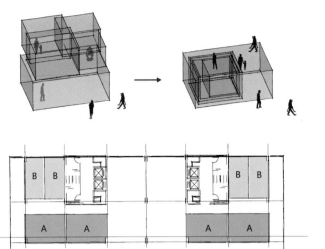

利用使用者的使用时间差实现空间利用效率最大化
使用者离开时私人空间收缩，公共空间扩展
有使用需求时私人空间扩展

Maximize space utilization efficiency by using the user's time difference
Private space shrinks when user leaves while public space expands
Private space expands when there is demand

提供家庭户型 A 和单人户型 B
A 户型可根据个人需求定制

Provide household type A and single type B
A type can be customized according to individual needs

单元变化示例
UNIT CHANGE EXAMPLE

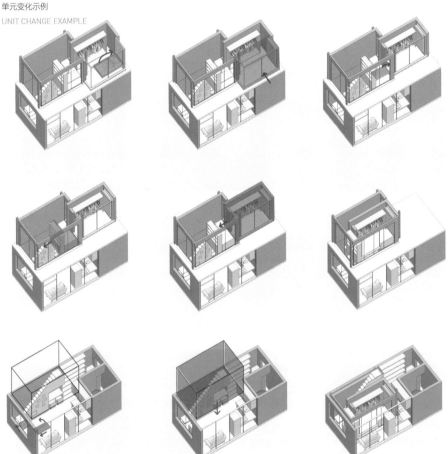

利用悬挑 3 米空间设置可移动的空间盒子，具有不同功能的空间盒子可体现"物联网"的便利以及"物——物"交换以及小型公共空间的共享，将原有"人到物"的关系，转变为物"FIL TO ME"的关系，符合现在人的生活模式，同时实现真正共享。

Use the cantilever 3m space to set up the movable space boxes. The space boxes with different functions can reflect the convenience of "Internet of Things" and the exchange of "object-object" as well as the sharing of small public space. The original "people-objects" relationship will be transformed into the "FIL TO ME" relationship of objects. This is in line with the current life mode of people and at the same time achieves true sharing.

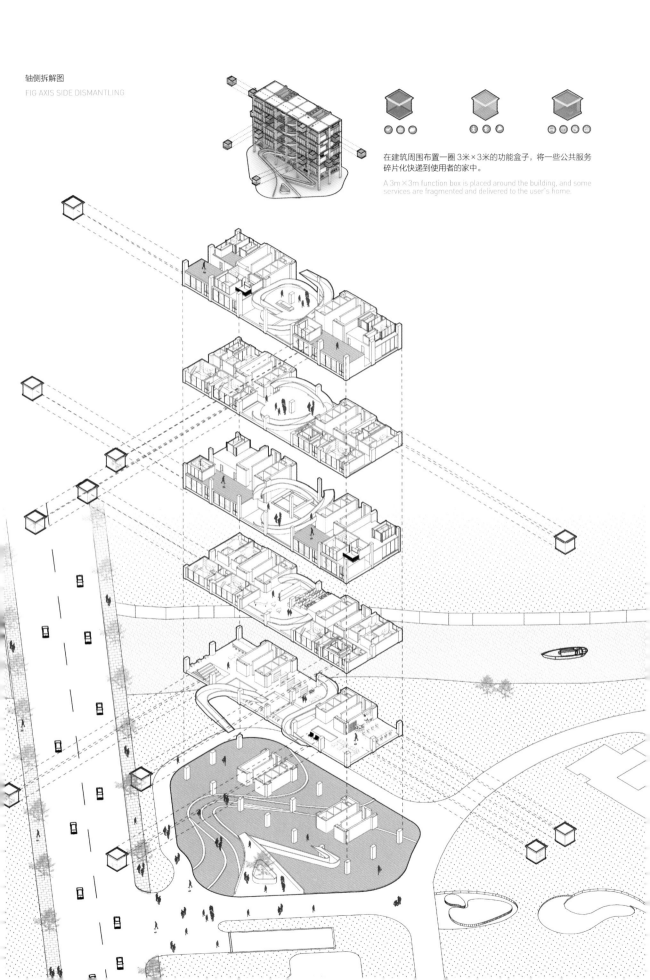

| 荣获奖项：入围奖 | Awarded: Finalist Award |

定制生活，共享社区
Private Customized Space & Shared Community

团队成员：姚潮锋，何小丹
来自：之间建筑设计（深圳）有限公司

Team members: Yao Chaofeng, He Xiaodan
From: Zhijian Architectural Design (Shenzhen) Co., Ltd.

居住单元
LIVING UNITS

随着经济与科技的发展，人们的生活越来越多样化，常规的居住单元往往满足不了个性化的需求。本设计通过智能移动隔断装置系统，自动实现个性化空间的定制，并具有灵活可变性。个性化空间衍生多样化的功能空间，使居住者之间的生活空间产生差异。居住者可以通过智能移动隔断装置系统控制墙体的开启与闭合，实现资源的共享，从而形成有活力有交流的社区。

With the rapid development of society and technology, living patterns have become more diverse. The conventional residential units often can not meet the individual needs. By our intelligent mobile partition system, personal space can be achieved automatically. The personalized space with flexibility enables a variety of functional space which allows the individual living space to be customized and different. By controlling the opening and closing of the wall of the intelligent mobile partition system, people can share resources and form a vibrant and social community.

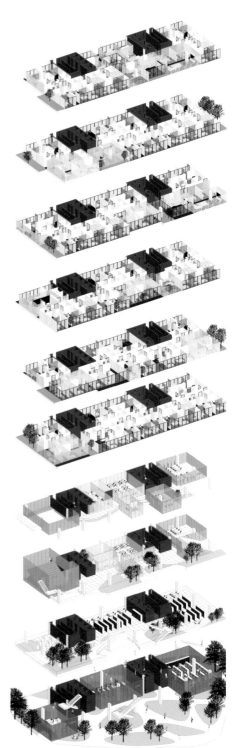

自动移动隔墙装置系统
AUTOMATIC MOBILE PARTITION SYSTEM

自动移动隔墙装置，轨道根据身体和家具尺寸，为1200mm间隔纵横排布。

The track arrangement of the intelligent mobile partition system is 1200mm crosswise according to body and furniture size.

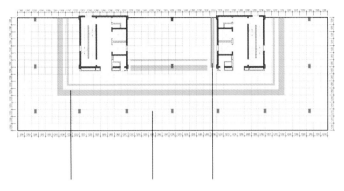
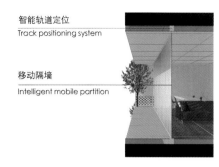

智能轨道定位
Track positioning system

移动隔墙
Intelligent mobile partition

主要配套空间及管线位置
The space of form a complete set

主要使用空间
Main space

移动隔墙预留存储空间
Intelligent mobile partition storage space

通过对功能的需求分析，形成个性的定制空间；通过分析其私密性，选择开放程度。
Through the demand analysis of the function, a personalized customized space is formed; by analyzing its privacy, the degree of openness is selected.

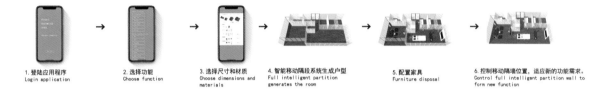

1. 登陆应用程序
Login application

2. 选择功能
Choose function

3. 选择尺寸和材质
Choose dimensions and materials

4. 智能移动隔段系统生成户型
Full intelligent partition generates the room

5. 配置家具
Furniture disposal

6. 控制移动隔墙位置，适应新的功能需求。
Control full intelligent partition wall to form new function

通过手机 APP 就可以选择和定制个人空间，包括尺寸，材质，及家具。并且可以时刻控制隔墙的开放与闭合，实现共享。
You can select and customize your personal space, including size, material, and furniture, through the mobile APP. And you can control the opening and closing of the partition wall to achieve sharing when.

节能
ENERGY-SAVING

夏季通过植物遮阳和格栅的自然通风，实现降温
Cooling is achieved by natural ventilation through plant shading and grille in summer

冬季通过双层表皮实现室内保温效果
In winter, indoor thermal insulation effect is achieved through double skin

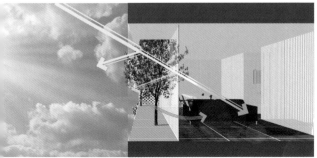

荣获奖项：入围奖　　　　　　　　　　Awarded: Finalist Award

宜居 +　　　　　　　　　　GREEN LIVING +

陈亚君、谭高飞（有行建筑工作室）、陈玉森博士（智能交通专家，荷兰国家应用科学院/TNO资深研究员，兼职 TU Delft 和北京工业大学教授，中国交通运输部研究院/国家智能交通技术及研究中心特聘研究员）、杨帆博士（鲁汶大学计算机学院，imec-DistriNet 课题组成员）

Team members: Chen Yajun, Tan Gaofei (Youxing Architecture Studio), Doctor Chen Yusen (expert in intelligent traffic, Senior Researcher of Netherlands Organisation for Applied Scientific Research/ TNO, part-time TU Delft and Professor of Beijing University of Technology, Distinguished Researcher of China Academy of Transportation Science/ National Intelligent Transport Systems Center of Engineering and Technology), and Doctor Yang Fan (member of imec-DistriNet Subject Group of College of Computer, Catholic University of Leuven)

都市农业和360度全景的概念可以作为提案"宜居+"的设计原动力。当下，青年人才白日繁忙的办公室工作，家仅仅是晚上睡觉的场所。互联网的便利，点餐、网购成为日常，而传统的买菜、做饭、打扫则无暇顾及。我们希望根据新的生活方式重新调整户型的使用构成。

The urban agriculture and 360° view become the initial power of the "Comfortable Living+". Nowadays, most young people work in the office and only sleep at home. They prefer ordering food or buying daily necessary on internet. They have much less time for cooking or cleaning the rooms. We hope that the apartment structure could be adjusted according to the new lifestyle.

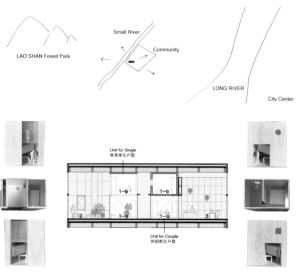

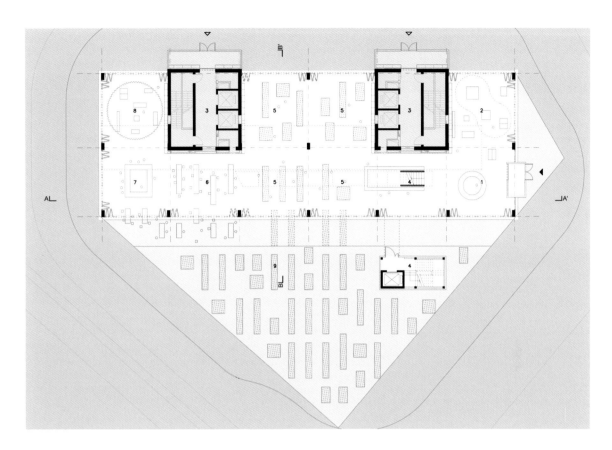

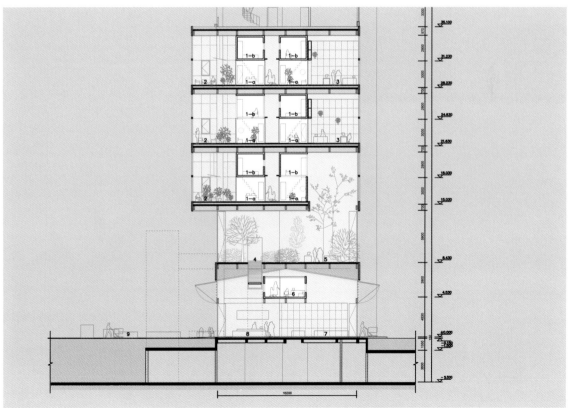

剖面图 Section B-B

Conventional Apartment

Shared Working / Meeting ......   Green Living + Apartment   Shared Kitchen / Dining / Living / Garden / Gym / Theatre ...

公共性

对于未来人才公寓的提案是合理紧凑地安排个人空间，提供更多的面积给共享、交流的楼层社区空间。楼层共享空间由大楼物业管理部门来维护，而居民们则可自由地使用不同的共享空间。绿植和都市农业是带来公共交流和竖向空间延伸的主线。

Publicity

Our proposal for the future talent apartment is compacting the individual space, offering surfaces for the sharing life. The property management will maintain the sharing space. The residents can freely use the diverse sharing rooms. Green and Urban agriculture is the clue to bring the public interaction and vertical space extension.

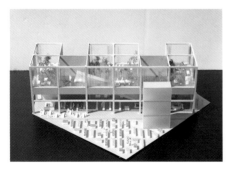

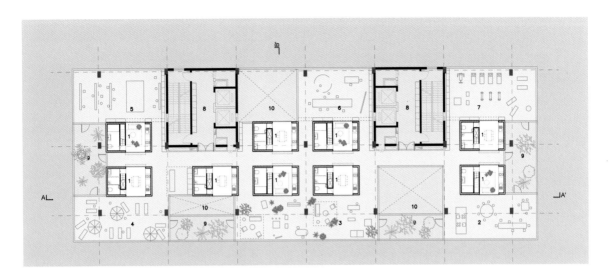

提供了 2 种基本的单元户型，分别给单身和伴侣使用。

There are 2 types of units, offered for singles and couples.

物联网 Internet of Things

物联网在集合的层面上收集、分析数据、而后执行，能发挥更大的效能。

Internet of Things will be more efficient in terms of data analysis and management in a collective manner.

环境平衡－水循环系统 Autarkic

环境平衡－绿"肺"及多层次保温隔热

Environment Balance-Water Circulation System Autarkic

Environment Balance - Green "Lung" and Layering Insulation

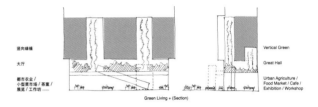

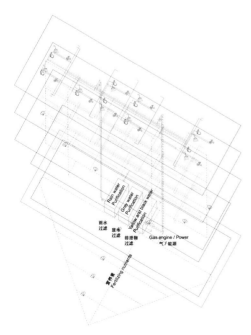

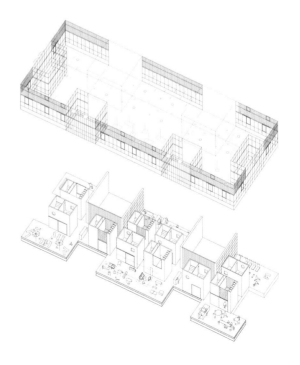

荣获奖项：入围奖　　　　　　　　Awarded: Finalist Award

垂直社区的未来生活
——青年创客住宅

Future Life in a Vertical Community
——Residential Design for Young Maker

团队成员：曾文涛
来自：中央美术学院

Team members: Zeng Wentao
From: China Central Academy of Fine Arts

对下一代建筑的理解
UNDERSTANDING OF THE NEXT ARCHITECTURE

唯有勇于面对、敢于尝试，才有可能从现实世界的无能处境中脱身，寻求一个未来的可能解答。下一代建筑在解决社会问题与满足人们需求的同时，亦应向着更加便捷、以人为本、经济环保型的方向前进。

Only by daring to face and to try can it be possible to escape from the incompetence of the real world and seek a possible solution for the future. While solving social problems and satisfying people's needs, the next architecture should also move in a more convenient, people-oriented, and economically environmentally friendly direction.

对垂直社区的理解
UNDERSTANDING OF VERTICAL COMMUNITIES

对垂直村落与社区的理解来自 MVRDV 对建筑自生长与运作的探索，在理解与引用之后，垂直社区生存与发展有以下几点主要因素：
1）社区密度
2）居民个体性
3）社区关键群众
4）建筑的弹性
5）集体性
6）进化式成长
7）社区多样化
8）人性尺度
9）社区公共性
10）社区与建筑的非正式性
——而这些种种便是设计垂直社区时我要着重考虑的设计因素。建筑始终应该从所在环境和使用者的需求出发，形式反而不是太重要。

The understanding of vertical villages and communities comes from MVRDV's exploration of the self-growth and operation of buildings. After understanding and quoting, vertical community survival and development have the following main factors:
1) Community density
2) Individuality of residents
3) The key people in the community
4) The elasticity of buildings
5) Collectivity
6) Evolutionary growth
7) Community diversification
8) Human scale
9) Community publicity
10) The informality of communities and buildings
—These are the design factors that I need to focus on when designing vertical communities. Buildings should always be based on the environment and the needs of users. The form is not too important.

核心理念阐述与创新
CORE CONCEPTS AND INNOVATION IN THE NEXT GENERATION OF BUILDINGS

本作品试图讨论演变中的建筑，年轻创客在城市中集结，通过现有场地和预制单体的结合，聚集模式利于他们更加频繁的人际交往及面对面交流。
1）通过打破传统的建筑模式，住宅的单体体量通过旋转，创造出各种使用及联系的可能性。大规模生产向数字化定制型社会演变。2）在场地原有结构上加上斜撑结构，用于加固单体体量的安装与拼接，并分隔出菱形空间。3）通过增加圆柱体墙面及楼板使空间合理运用，管道及储藏空间藏于结构空间中。4）切分楼板及增加空间体量，丰富空间层次；交通空间及公共空间＆绿化的介入，使单体模型趋于完整。5）挂壁式结构利于预制模块的安装，大小不同的体量满足不同的功能需求。安装以组团、叠加的方式进行，满足居民的组织兼容性与真正的空间共享。6）居住者的需求丰富了交流空间和社区中心，媒体立面的使用增加了人与人之间的互动和交流。7）新的建筑空间形式带来新的用户体验，通过机械升降装置将藏于底部的物品抬升来满足紧张的空间需求。8）信息储藏成为未来信息化社会的关键，合理利用空间利于提高效率及空间使用率，轴承装置将储物空间扩大。9）管道及连接装置隐藏于特殊空间，单体组装完成后参与整体建筑的生态循环中去。

I tried to discuss the evolving architecture. Young creators gather in the city. Through the combination of existing venues and prefabricated monomers, the aggregation model facilitates their more frequent interpersonal communication and face-to-face communication.
1) By breaking the traditional construction model, the amount of singletons in the house creates various possibilities for use and connection through rotation. Large-scale production evolves into a digitally customized society.
2) The diagonal brace structure is added to the original structure of the site to reinforce the installation and combination of the monomer volume, and separate the diamond space.
3) By adding cylinder walls and floor panels to make the space to be reasonably used, pipelines and storage space are hidden in the structural space.
4) Cut the floor and increase the space volume, and enrich the space level; interfere in traffic space and public space & greening, so that the monomer model tends to be complete.
5) Wall-mounted structure facilitates the installation of prefabricated modules, and different body sizes meet different functional requirements. The installation is carried out in a group and superposition manner to meet the organizational compatibility and real space sharing of residents.
6) Communication space and community centers are enriched by the needs of residents, and the use of media facade increases the interaction and communication between people.
7) The new form of building space brings new user experience, lifting the objects hidden in the bottom by the mechanical lifting device to meet the tight space requirements.
8) Information storage will become the key to the future information society. Rational use of space will help improve efficiency and space utilization, and bearing devices will expand the storage space.
9) Pipelines and connecting devices are hidden in special spaces and participate in the ecological cycle of the whole building after the individual assembly is completed.

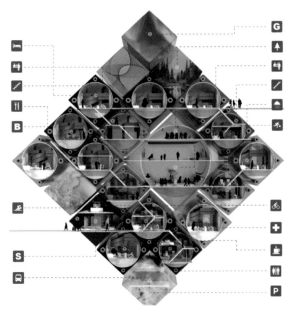

| 荣获奖项：入围奖 | Awarded: Finalist Award |

UPLIFE/ 共享生活：住宅之外更是空间乃至全身心的改变

UPLIFE/Shared Life: Beyond Dwelling, Transforming Spaces, and even Minds and Hearts

团队成员：Sinan Gunay，Nurhayat OZ
来自：土耳其 SuperSpace 公司

Team members: Sinan Gunay, Nurhayat OZ
From: SUPERSPACE, Turkey

Uplife 建筑可以作为微观解决方案为未来提供借鉴，它用最小的改动模糊了公共和私人空间的界限，创造了一个宜居的社区，增强了垂直公共领域的功能如城市农场和花园，高效节能地对这些公共领域中的生产、养殖和废物进行处理，并运用窄带物联网监控、控制和分析可再生能源，打造一个自给自足的闭环。

This building can be delivered as a micro-scale solution to what comes next. with minimal destruction that blurs the boundaries of the public and the private, a liveable community is fostered, enhancing the vertical public realm, urban farms and gardens are spread out in the existing structure which harbours producing, consuming and waste in circulation with energy efficiency and renewable energy monitored, controlled and analysed with narrowband IoT, creating a closed loop of self-sufficient life.

一切以人为中心试验性住宅区
IT'S ALL ABOUT PEOPLE
EXPERIMENTAL RESIDENCE UNITS

我们爱自己的家，但我们永远不想回去。我们爱在外闯荡，但我们也永远无法离开家。这就是人性。在充满压力的都市生活中，人很容易无缘无故地感到压抑。因此，我们将住宅区和公共社交区融为一体，缩短人与人之间的距离，模糊私人和公共区域的边界，让人们更容易适应现实的世界。

We love our homes but we never want to get in, we love going outside but we can never get out of home. It is the human nature. It is easy to get depressed out of nothing in stressfull lives of urban cities. Integrated residential and social public spaces make it easier to adapt the real world by shortening the transition between eachother and/or blurring the boundaries of the private and the public zones.

水平和垂直农业
HORIZONTAL AND VERTICAL FARMING

提倡步行、交谈、倾听、冥想、分享、工作、种植、园艺、购物、烹饪……过更健康的生活……各个层面公共社交空间都有着更高的渗透性和可见性，Uplife为邻里和农民开发了手机应用程序，日程表简单易遵循，并鼓励大家互相帮助，从身心层面建设一个更加健康的社区。

Promote walking, talking, listening, meditating, sharing, working, farming, gardening, shopping, cooking ... to live a healthier life ... increased permeability and visibility in public social spaces on all levels, UPLIFE phone applications for neighbours and farmers making it easy to follow a calendar and promote helping each other fostering healthier communities both physical and psychological.

共享公共空间 垂直农业
SHARED PUBLIC SPACE VERTICAL FARMING

该混合项目能够统筹协作和个性化，共享农业空间可以通过共同任务和奖励措施来建立社区。选择本地作物物种来营造一个自然且可持续的生活环境。

Hybrid approaches combining aspects of collaboration and individualization, shared farming space can help build communities around shared tasks and rewards. Local plant species to be selected in order to provide a natural context and a sustainable habitat network.

| 荣获奖项：入围奖 | Awarded: Finalist Award |

塔罗社区　　　　　　　　　　　　TARRO Community

团队成员：陈丹丹，黄厚渝，尹正，赵航，张贝贝，赵睿智
来自：华侨大学

Team members: Chen Dandan, Huang Houyu, Yin Zheng, Zhao Hang, Zhang Beibei, Zhao Ruizhi
From: Huaqiao University

设计说明
DESIGN NOTESR

从人、文、产、地四个要素切入，通过实地调研，总结出在居住、环境、业态、生活等方面有待解决的矛盾。在此基础上，结合 VR，物联网等前沿科技，运用垂直互联的公共空间，推拉式模块的居住空间，迎合未来人居生活方式。塔——高耸的塔形建筑。为解决人口密度大，自然资源少等问题，现代居住建筑多垂直发展，习塔以构，似塔而筑。塔罗——西方占卜，偏向于心灵占卜。"占"不同人的不同行为心理，"卜"算细微可见预期。其身之所屈，即为心之所居。物欲社会下灵魂居室，内心与外界的交互联通，如千变塔罗，考验策作之匠心。用以推拉模块，契合未来生活理想。

Through field research, sum up the contradictions in the living, environment, business, life and other aspects to be resolved. On this basis, combined with VR, Internet of Things and other cutting-edge technologies, use vertical interconnected public space, push-pull module living space, to cater to the future living lifestyle. Tower-- high-rise tower-shaped buildings. In order to solve the problems of large population density and fewer natural resources, modern residential buildings are developed vertically, and the towers are constructed like towers. Tarot - the West dominates, leaning toward the soul. "Accounting" different people's different behavioral psychology, "Bu" is slightly visible expectations. The grievances of the body are the home of the heart. In the materialistic society, the soul room is located, and the inner and outer interactions, such as the ever-changing Tarot, test the ingenuity of the plan. It is used to push and pull modules to meet the ideals of future life.

单体设计说明
MONOMER DESIGN NOTES

公共区域
PUBLIC AREA

采用竖向混合的应对策略，加强各个功能之前的自然连接，打破孤立的办公娱乐工作模式，加强竖向空间的自然过渡，使空间更加多元。

Adopt vertically mixed coping strategies to strengthen the natural connection before each function, break the isolated office entertainment work mode, and strengthen the natural transition of vertical space to make the space more diverse.

居住单体模块
RESIDENTIAL UNIT MODULE

在每一个居住单元植入随型易变的推拉盒子，在满足居住需求的同时，使住户有更加多的选择方式对自己的空间进行定义。与此同时，打破传统的居住模式，加强在居住空间中的公共空间，从而使邻里之间更多的互动与交流。

In each living unit, a portable push-pull box is inserted to meet the living needs, and the occupants have more choices to define their own space. At the same time, it breaks the traditional living mode and strengthens the public space in the living space, so that more interaction and communication between the neighborhoods can be obtained.

细部说明
DETAIL DESCRIPTION

墙体采用中空夹层设计，外部可推拉墙体嵌套其中，可通过上下各五道滑轨进行推拉移动，并承载其自重，以达到随型易变的空间变换效果。

The wall adopts a hollow sandwich design, and the external push-pull wall is nested therein, and can be pushed and pulled by the upper and lower five slide rails, and carries its own weight, so as to achieve a spatially variable effect with the type.

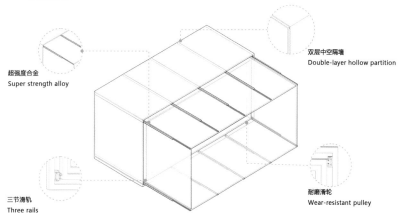

超强度合金
Super strength alloy

双层中空隔墙
Double-layer hollow partition

三节滑轨
Three rails

耐磨滑轮
Wear-resistant pulley

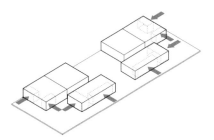

模块推拉前示意图
Module push-pull schematic (Before)

推拉前楼层公共空间及视觉空间形态示意图
Schematic diagram of public space and visual space on the floor before pushing and pulling

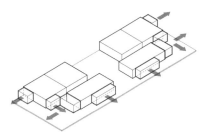

模块推拉后示意图
Module push-pull schematic (After)

推拉后楼层公共空间及视觉空间形态示意图
Schematic diagram of public space and visual space on the floor after pushing and pulling

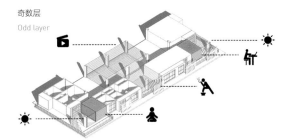

推拉前楼层公共空间使用功能示意图——由更为私密的休憩空间转变为更大尺度的聚会空间
Function of using the floor public space before the push-pull——From a more intimate open space to a larger scale gathering space

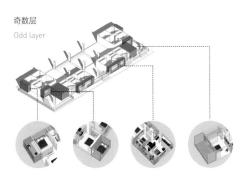
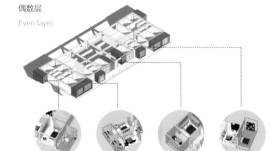

推拉后楼层室内空间使用功能示意图——根据功能需求拓展出卧室、阳台等使用空间
Function of using the floor interior space after the push-pull——Expand the use space of bedrooms, balconies, etc. according to functional requirements

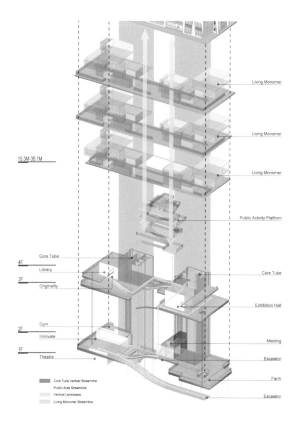

流线分析图
Streamline Analysis Chart

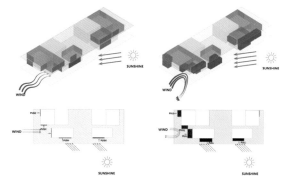

居住单元推拉模块对建筑热环境影响示意图
SCHEMATIC DIAGRAM OF THE IMPACT OF THE RESIDENTIAL UNIT PUSH-PULL MODULE ON THE BUILDING THERMAL ENVIRONMENT

夏季
当体块归于原位时，体块之间形成空隙，有利于风吹入建筑内，降低建筑内部温度。与此同时，朝阳面体块归于原位时，所受阳光照射较弱，降低室内温度。
Summer
When the block is in situ, a gap is formed between the blocks, which is beneficial to the wind blowing into the building and reducing the internal temperature of the building. At the same time, when the sun block is returned to the original position, the sunlight is weakened, and the indoor temperature is lowered.

冬季
当体块抽拉出时，体块之间进行封闭式组合，阻挡风进入建筑内从而对建筑有保温作用。与此同时，朝阳面体块抽拉出时，所受阳光照射较强，使室内温度升高。
Winter
When the body block is pulled out, the body blocks are closely combined to block the wind from entering the building and thus insulating the building. At the same time, when the sun block is pulled out, the sunlight is strongly exposed, and the indoor temperature is raised.

荣获奖项：入围奖　　　　Awarded: Finalist Award

城中游牧：移居与一体化生活

Nomadic City: Integrative Life of Movability

团队成员：刘志军，张兵、黄恩泽、张安强、王琳嫣、徐延峰
来自：江苏省建筑设计研究院有限公司

Team members: Liu Zhijun, Zhang Bing, Huang Enze, Zhang Anqiang, Wang Linyan, Xu Yanfeng
From: Jiangsu Architectural Design Research Institute Co., Ltd.

Variational Activity

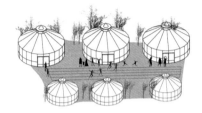

Variational Activity

Various Combination

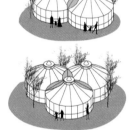

Classical Combinition

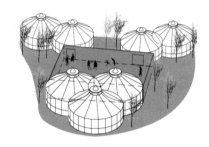

Auxiliary Function

对下一代建筑的理解：一成不变的居住空间难以满足未来生活的需求，因此预制化与集约化相结合可以作为未来住宅建筑的发展方向。空间的共享和变化提供了更多的选择，同时给人们带来了舒适感。如今，考虑到住房价格，家庭组成，人工智能和物联网，大城市的人们在寻找舒适、有趣、互动和绿色的生活空间，就像游牧民族的生活一样。

Understanding of the Next Architecture: Stable residential space can hardly meet the future development needs, so the combination of prefabrication and intensification can be used as the development direction of future residential buildings. Clearly, the sharing and variability of space provide more choices while giving people comfort.Nowadays, considering housing price, family composition, artificial intelligence and Internet of Things, people,especially in large city, are looking for comfortable, funny, interactive and green living space, just like the life of Nomads.

设计概念：为改变固定、单一、公共与居住空间隔绝的现状，创造一种可变、共享、回应气候变化的游牧生活，设计以单元模块移动、组合的方式在垂直与水平两个方向上形成集居住、公共活动、绿化于一体的良好生活多样、动态的生活体验。

Design Strategy: In order to change the status quo of fixed, single, isolated condition of public and residential space, it is necessary to create a nomadic life that is variable, shared and responding to climate change. The design uses the unit module to move and combine so as to form a good living and dynamic life experience integrating living, public activities and greening in both vertical and horizontal directions.

SMARTCELL
Concept

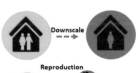

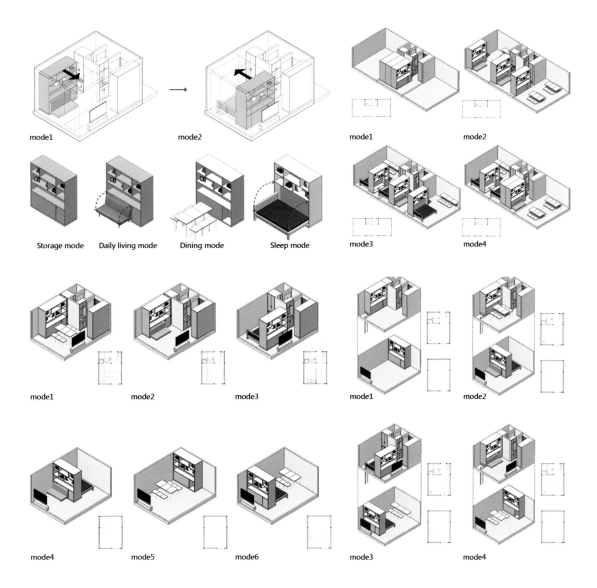

居住单元的移动组合：居住单元以3600毫米×4900毫米的尺寸为基本模块，模块间的移动组合可以形成单居室、双居室等多种户型，满足未来变化中的居住需求，同时，公共空间随居住单元的移动变化呈现多样化的形态。居住单元内部采取可变可移动的家具，提高了有限面积的使用效率。

Mobile Combination of Living Units: The living unit is based on the size of 3600mm×4900mm. The mobile combination between modules can form a single room, two-bedroom and other types of houses to meet the changing needs of the future. At the same time, the shape of public space varies with the movement of the living unit. The movable unit inside the living unit adopts variable movable furniture, which improves the use efficiency of a limited area.

沈迪
Shen Di

全国工程勘察设计大师，教授级高级工程师
华建集团副总裁兼总建筑师
中国建筑学会建筑师分会副理事长

National Master of Engineering Survey and Design, Professor-ranked Senior Engineer
the Vice President of Arcplus Group and Chief Architect, Vice President of the China Architectural Society of Architects

科技重塑邻里社区
Technologies Reshape Neighborhoods and Communities

如今，科技已影响到居住与生活的各个方面，比如出行、购物等等。未来，科技会更进一步与生活融合，而非简单地影响生活。可以说，随着技术迅速迭代，科技将通过对建筑与社区的重构，全方位影响我们的居住及生活环境。建筑发展趋势，将根据不同的建筑类型，走向不同的未来。针对城市居住类建筑，必须关注三个方面，第一是舒适性，提升居住质量依然是住宅建筑首要的发展目标；第二是绿色环保，在有限的资源环境下，建筑必须符合可持续的时代发展要求；第三是安全性，在高密度的城市垂直社区中，对于安全的需求尤为突出。

Technologies are now influencing various aspects of people's life, including living, traveling and shopping. In the future, technologies would further integrate with our life. It's fair to say that fast updating technologies will influence all aspects of our living environment by restructuring our communities. Development trends vary for different types of buildings. For urban residential buildings, we must attach importance to three aspects. The first is comfort. The primary development goal for residential buildings is still improving living standard. The second aspect is environmental protection. With limited environmental resources, architecture must be in line with sustainable development standards of the new era. The third is security. Security is vital for vertical communities of urban built-up areas.

以"下一代建筑"为议题的此次竞赛，涉及未来居住建筑的两大核心问题：首先是高密度城市社区及公共资源的重建与整合。目前，中国城市已经越来越向垂直发展，超高的建筑容积率、建筑覆盖率及人口密度，日益改变了建筑中的人居状态。高楼林立的住宅与小区，并未找到社区关系一种很好的建筑解决方式，邻里间往往十分生疏，甚至完全没有往来。这种生活状态与中国传统的居住方式相差甚远。如何进行邻里关系的重建，如何将技术和建筑设计很好地结合起来，让原有的生活与文化传统在新的环境中再生，是我们面临的比较大的问题。

Themed on "Next Architecture", this competition involves two key issues of future residential buildings. First is the reconstruction and consolidation of urban built-up areas. At present, Chinese cities tend to grow vertically. High floor area ratio, building density and population density, all have changed people's living conditions inside buildings. Residential areas with high-rise buildings are not the best way to enhance people's relations within the communities. On the contrary, neighbors become unfamiliar and even never get in touch with each other. This is far different from traditional Chinese way of living. Now we are faced with major questions like how to rebuild relations between neighbors, how to well combine technologies with architecture design, and how to revitalize former way of living and cultural traditions in the new environment.

另一个核心问题是：随着如今互联网技术、AI 技术等各项技术的飞速发展，如何把这些前沿技术融入居住建筑设计中，进而共同服务于生活需求？这也是当下建筑界面临的重大挑战。众所周知，我们正在居住建筑设计中逐步推广各种信息化与绿色节能技术，但同时也面临了很多问题，比如很多建筑是设计完了方案，再考虑把信息与节能技术加入进去，两者之间缺乏交流，有时就成了技术的堆积，也导致新科技与建筑设计实际上的脱节。

Against this backdrop of rapid development of Internet technology and AI technology among others at present, another vital question is that how we could integrate these cutting-edge technologies into the design of residential buildings and thus meet people's different needs in life. This is one of the major challenges the building industry face today. It is widely known that we are stepping up our effort to integrating various information technologies and green technologies in residential architecture design. Yet there has still emerged many problems. For example, it is common that one may consider integrating information and energy saving technologies into the building after the design is finished. Lacking in communication between the two phases may sometimes leads to overlapping in technologies and inconsistency between design plan and new technologies to be applied.

在高密度的城市，如何建立起好的邻里关系，如何让高密度与绿色空间兼得，如何重建社区生活，是下一代居住建筑必须重点考虑的议题。此次智慧树竞赛项目关于"垂直社区未来生活"的主题，本身就直接涉及这一问题。其中的参赛作品更是亮点纷呈，对于亚洲垂直社区的营建，表达了很多先进的理念与设计方法。与此同时，由于鼓励科技企业、大数据企业、科研机构等联合建筑师共同参与比赛，也体现出设计统筹、协同创新的过程，作品更是对当代科技与传统建筑手段的渗透交融有很多尝试性的探索。这是一项很有意义的工作，建议对竞赛作品与得奖者以类似巡展的方式做进一步宣传，因为这不仅有助于这一项目本身的实施，也对于我们未来的居住建筑理念发展具有积极的推动作用。

While designing architectures for the next generation, we should consider key issues as follows: how to enhance relations between neighbors in built-up cities, how to strike a balance between building density and green space, and how to rebuild living environment in the community. Smart Tree competition's theme "Future Life in Vertical Communities" is directly related to these questions. Participants' remarkable entries have demonstrated many advanced ideas and design methods in building vertical communities in Asia. Meanwhile, we encourage architects to cooperate with technological companies, big data companies, scientific and research institutions among others to participate in this competition. This well reflects the concept of design planning and coordinated innovation. Moreover, participants' solutions tap into possibilities in combining technologies and traditional architecture skills, which is very meaningful. I suggest that we could hold traveling exhibition or take other similar measures to further publicize solutions in this competition and prize winners. This is not only helpful for the project implementation, but also promotes the development of residential building concept in the future.

冯正功
Feng Zhenggong

江苏省设计大师，中衡设计集团股份有限公司董事长、总建筑师
Design Master of Jiangsu Province. Chairman of the Board and Chief Architect of Arts Group Co., Ltd..

在延续中守护未来
Protecting the Future in Continuation

建筑是文化的切片，传统的积淀。作为城市的灵魂，它更是一部延续的历史。透过建筑，我们看到了历史，听到了故事，浸染了文化，了解了传统；透过建筑，历史、故事、记忆、文化、传统、环境得以延续。延续是一个空间概念，更是一个时间概念。延续的落脚点不是在过去，更是在当代或现代，甚至未来。吴良镛先生说："规划设计者不仅要有空间的自觉性，还要有时间的自觉性。要善于领悟过去，把握现在，预见将来。"通过延续，我们赋予建筑以现代功能、现代空间和现代体验；通过延续，建筑有了活力和新的生命，延续的价值才得以真正实现。生命在于延续，而生命的活力更在于延续加创新，秉承延续建筑这一理念，未来的建筑发展将会更多元化、个性化、人性化、生态化。在物理属性上，下一代建筑是生态自然与科技创新的结合体。在生命属性上，下一代建筑是地域土壤与基因延续的结合体。

Buildings are the slices of culture and the sedimentary deposits of traditions. As the soul of a city, buildings are even a continuing history. Through buildings, we have seen history, heard stories, been imbued with culture, and understood traditions; and through buildings, history, stories, memories, culture, traditions and environments have been continued. Continuation is a concept of space, and even a concept of time. The foothold of continuation is not only in the past, but also in the contemporary or modern times, and even in the future. Mr. Wu Liangyong has ever said: "Planning designers should have not only space consciousness, but also time consciousness, and should do well in understanding the past, grasping the present, and foreseeing the future." Through continuation, we endow buildings with modern functions, modern space and modern experience; and through continuation, buildings have vitality and new life, and the value of continuation can be truly realized.

某种意义上，延续是一种需求，是对历史、文化、传统、自然和环境的回应，更是对人需求的尊重，人性的回归。延续应当是广义的，不仅包括优秀的传统文化，还应包括记忆、场景、地域、地景、思想、故事等。科技与信息化的发展，正越来越改变人的生活与工作方式，而建筑与科技结合的主题将始终延续"以人为本"这一核心诉求。未来，数字化、参数化的介入将为新创意、新技术、新材料在建筑上的运用提供可靠的支持，使建筑在舒适度、安全性、智能化以及能源自给、绿色环保等方面获得全方位的突破。

Life lies in continuation, and the vitality of life lies in the continuation and innovation. Adhering to the concept of continuation of buildings, future buildings will be developed to be more diversified, individualized, humanized and ecological. In terms of physical attributes, the next-generation buildings will be the combination of ecological nature and technological innovation. In terms of life attributes, the next-generation buildings will be the combination of regional soil and genetic continuation.

In a sense, continuation is a kind of demand, a response to history, culture, traditions, nature and environments, and even the respect for human needs and the return of humanity. Continuation should be broad, including not only excellent traditional culture, but also memories, scenarios, regions, landscapes, thoughts, stories, etc. The development of science & technology and information technology is increasingly changing people's life and working mode, and the theme of the combination between buildings and technology will always continue the core demand of "being people-oriented". In future, the intervening of digitalization and parameterization will provide reliable support for the use of new ideas, new technologies, and new materials in buildings, and make buildings realize overall breakthroughs in terms of comfort, security, intelligentization, energy self-supply, green and environmental protection, etc.

如今，中国正在按照高质量发展的要求，大力推进绿色建筑与创新发展，不断提升人居环境品质，在"智慧树下一代建筑"竞赛这一项目中，我看到了很多有创意并有技术创新的作品。万物皆生命。未来，如何赋予建筑生命力，形成真正革新的下一代建筑，还需要每个建筑师带着历史责任感，与时俱进、砥砺前行。

Nowadays, China is vigorously promoting green buildings and innovative development, and constantly enhancing the quality of human settlement according to the requirements of high-quality development. In the project of "Smart Tree Next-Architecture" Contest, I have seen many creative and technically innovative works. Everything is alive. In future, we architects should keep pace with the time and forge ahead to make clear how to endow buildings with vitality and form truly innovative next-generation buildings with a sense of historical responsibility.

Daniel V. Hayden

丹麦 DISSING+WEITLING 建筑事务所合伙人，建筑总监
Partner and Director of Architecture Affairs of DISSING+WEITLING

体验创造终极未来
Experience Creating the Ultimate Future

建筑师首先应该拥有自己的信仰体系，他不只是空间打造者，更是体验缔造者。当建筑既是数据的入口，又是数据沉淀和分享的平台，当房屋具备了智能化趋势，当数据在人、建筑和应用之间循环累进……当所有技术的不同组成部分，对于建筑而言都开始具有异乎寻常的意义时，我们必须思考的是，如何能够使建筑和科技有机结合，从而在未来为人们创造更好的体验。

Architects should first of all possess their own system of beliefs, as not only do they make space, they also create experience. Architecture is not only an entry point for data, but also a platform for depositing and sharing data, especially when buildings possess the trend of intellectualization and data progressively circulates among people, architecture and applications. So of course, when the idea of 'Next Architecture' is introduced, all of the different components of it seem to make a lot of sense to us. We must therefore consider how we can work with different technologies while combining architecture in an organic manner in order to create a better experience for people in the future.

建筑发展的瓶颈往往在于：在建筑未建成时，如果没有亲自进入环境，很难判断和衡量最终的体验感。如今的 VR 和 AR 技术给了我们模拟环境的可能，然而，就算利用虚拟现实技术，我们依然很难分辨其有效性。通常，我们只能测量已经发生或正在发生的各项建筑数据，甚至利用最新的现代科技，也很难评估将来还未发生的体验感。因此，这次竞赛给建筑师一次研究未来建筑真实案例的机会，是令人激动的举措。建筑师可以尝试各种新想法、对技术交互和人居体验进行测试，随之得到运行的结果，即各项体验值数据库，以供综合分析进一步解决问题。计算机算法成为帮助建筑师在建筑中阐述思想的工具。

The difficulty with architectural development is that it is challenging to sort of judge and measure a building that is not yet built without having the experience of entering the environment yourself. So even with today's VR and AR technologies giving us the possibility to simulate the environment, it's still really difficult to distinguish its effectiveness and know if it will really function. Generally, we can only accurately measure things that have happened in the past, or are in the present. It's very difficult to assess, even with the latest modern technologies to predict how we will experience something when it is not yet occurred, something we can only measure in the future. Therefore, this competition gives architects an opportunity to study real-life examples of future buildings, which is an exciting move. Architects can carry out experiment with the latest ideas, test technical interactions and human experience whilst the result of the operation, namely, the database measuring the value of experience, can be provided for comprehensive analysis to further find solutions to problems. Architects can experiment with the use tools such as computerized algorithms to help formulate their ideas in architectural terms.

此次竞赛还尝试将垂直生活和智能理念融入项目中，由于涉及数据驱动设计、人工智能化、信息交互技术等众多新科技，我们要在这些框架内同时处理多项技术与建筑的融合是一次艰巨的任务。看完竞赛作品后，我也看到了不同参赛者的困惑。很明显，有的项目专注于人的体验感，其他的一些项目则是被科技理念带动的。而我认为，科技是一方面，但作为建筑师和设计师，最终必须思考人的体验感。

The competition also attempts to integrate the idea of 'Vertical life' and 'Smart tree' concepts into the project. This has become a difficult task for us to deal with as artificial intelligence along with information interaction technology and data-driven design are some of the many new technologies involved and we must integrate multiple technologies with architecture within these frameworks simultaneously. After reviewing the competition works, I noted the wide variety of solutions from the different competitors. Some of the projects focused mainly on the human experience, while others were clearly driven by technological concepts. In my opinion, technology is only one aspect of architectural design, it is through the eyes of architects and designers, where we must ultimately consider architectural solutions within the realm of human experience.

下一代建筑，意味着建筑行业一次跨越式的发展。建筑学是否会被诠释成为一种能自身不断学习和持续进化的数据生态，并依靠数据技术在模拟应用场景中的探索、分析与回应，使设计师获得实时反馈，从而更好地与团队合作，让以往静态的建筑设计过程演变为动态交互式的建筑设计状态。

The 'Next Architecture' is a great advancement in architectural development within the construction industry. Designers will be able to get real-time feedback as well as explore, analyze and respond to simulated application scenarios when architecture is interpreted as living data that continuously learns and evolves itself. Therefore, it is important for this project to start cooperating with the team so that the previous architectural design process can evolve into a more dynamic and interactive architectural design process.

无论在何种情况下，对于建筑设计师、学生或竞赛者，依然需要明确：建筑设计师要通过人的体验感来考虑建筑设计，在下一代建筑设计上我们需要思考的是如何将各项科技融合在一起，服务于人的感官和需要。自始至终，体验都是未来建筑设计的终极角色。

In any case, it needs to be clear that whether an architect, student or competitor, architectural design should be considered through human experience. In 'Next Architecture', which is the next generation of architectural design, we need to consider how to integrate various technologies in order to meet the needs and demands of people. From the start to finish, the ultimate role of the upcoming architectural design is in fact working to improve the experience of architecture.

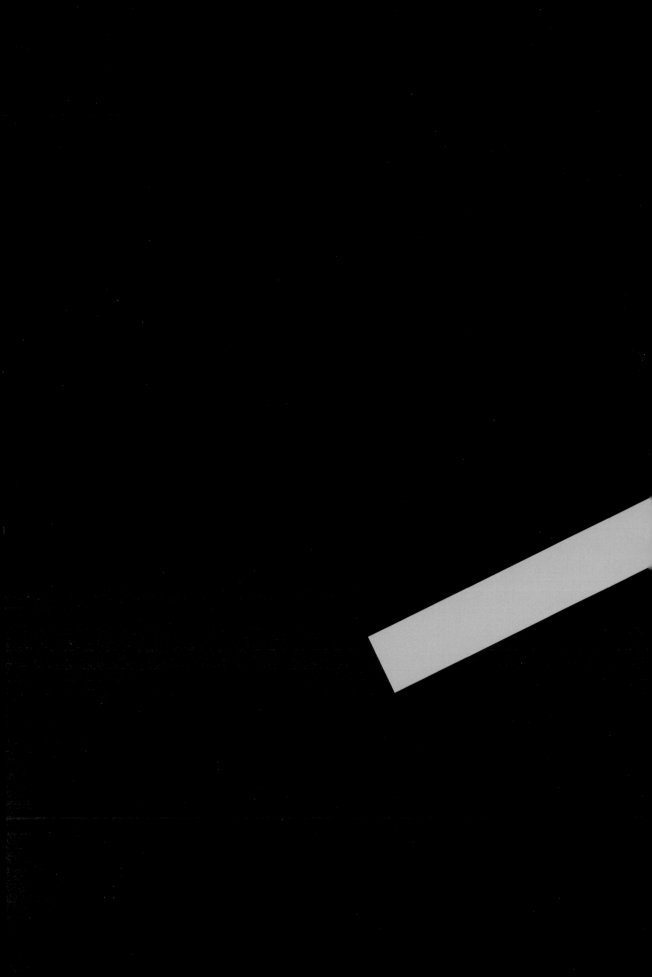

建筑科技应用篇

Chapter of Architectural Technology Application

荣获奖项：模块奖　　Awarded: Unit Module Prize

Verti- Community

团队成员：徐佳楠、张祺媛、韦柳熹、曾婧如、刘宁琳
来自：东南大学

Team members: Xu Jianan, Zhang Qiyuan, Wei Liuxi, Zeng Jingru, Liu Ninglin
From: Southeast University

A. "智慧树"设计（产品）定位？
基地周围多为集中式的传统小区和人才公寓。周边缺少公共空间及办公、商业、生活服务空间。即使有少量服务空间，只服务于组团内部，人群聚集性差。因此我们提出智慧树的总体定位：居住空间＋配套公共空间＝混合型功能空间。

A. What is the "Smart Tree" product positioning?
The site is mostly surrounded by centralized traditional residential communities and talent apartments. There are few public spaces and offices, business and service spaces, such as markets around the site. Even if there are a small number of service spaces, such as stores, it only serves the inside of their communities, and their extroversion is poor. Therefore, we propose the general orientation of the "Smart Tree ". This includes residential space along with supporting public space to result in a mixed functional space.

B. "智慧树"为未来人群提供什么？
我们希望智慧树未来服务的人群有三类：青年、创客、原住民。对青年提供最低成本的居住空间＋最丰富的就业平台；对创客 24 小时的居住、办公、餐饮空间，满足其高效生活的需求；对原住民提供丰富的公共空间。

B. What does the "Smart Tree" design provide for the future population?
We want the "Smart Tree" design to serve three groups of people in the future: young people; entrepreneurs; and the natives. Provided for the youth: Minimum Cost of Living + Multiple Opportunities for Employment; for entrepreneurs: 24-hour spaces for living, working and eating, which suffices the high demand for efficiency; for local residents: the abundance of public space will be provided on site.

C. 智慧树如何吸引未来的人群？
C. How does the "Smart Tree" design attract people in the future?

高效 + 开放：以前，传统小区居住形式为点状集中式住宅。组团内部只解决居住需求；传统小区为垂直生长的居住空间。办公、商业、生活空间在组团外外水平发展；小区的封闭性和水平维度交通的可达性造成了出行的低效。现在，"智慧树"以垂直社区方式打破传统小区的水平低效，提供丰富的功能空间，实现衣食住行的高效性。

Efficiency + Transparency: Past: The traditional residential communities were a kind of point-like centralized housing. The group only addressed the residential needs. The traditional residential communities are vertical residential spaces. Offices, businesses and residential spaces are outside of the groups. The closure of the community and the inaccessibility of horizontal traffic have made getting around inefficient. Now: The "Smart Tree" vertical communities break the low efficiency of traditional communities. They provide an abundance of functional spaces which makes sure the efficiency of our daily life is met such as food, clothing, housing and transportation.

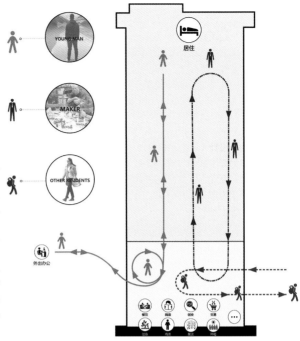

健康的生活方式：以"打造最健康的生活方式"为理念，在所有户型中设置微小农场。种植的蔬菜可以自食，也可售卖。室外广场的蔬菜种植区延伸到室内公共空间，成为可食地景的同时，也提供了公共空间内销售的产品。

Healthy Lifestyle: With the concept of "building the healthiest lifestyle", we set up tiny farms in all types of living units. Vegetables produced here can be eaten as well as sold. The vegetable planting outdoor area also extends to the indoor public space, which becomes an edible landscape and also provides products for sale in the public space.

智能：居住空间模块基于目标编程设计，运营模式引入物联网运营。

Intelligence: Residential space modules are designed based on objective programming while utilizing the 'Internet of Things' into the operation.

节能技术：外墙面种植绿植，雨水灌溉绿植后多余的水分和室内废水通过每一户型的水循环系统实现自维持。

Energy-saving Technology: Green plants are planted on the exterior walls, and the excess water from rainwater after irrigation and indoor wastewater is then self-sustained through the water circulation system of each household.

空间的灵活多变：通过空间分隔及家具组合形成多功能组合变化的空间。

Flexibility of Space: The multi-functional space is formed from a combination of a variety of changes of space and furniture separation and assembly.

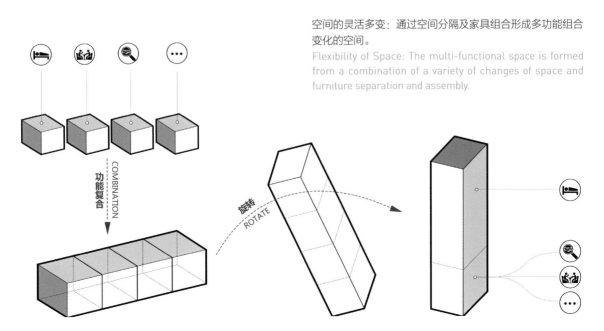

D. 智慧在何处？
D. Where is the Intelligence?

编程设计：由数字化设计根据限制条件进行模块化居住空间生成，灵活高效。根据避让点的选择，智能调整建筑形式。
Programming Design: The modular residential space is generated by just one-click, and the building form is intelligently adjusted according to the selection of the avoidance points which can make the results efficient, flexible and environmentally friendly.

云交互：通过 APP 的开发和智能交互装置实时感应收集"智慧树"中活动的大数据，形成基于网络数据平台及线上交流论坛。
Cloud Interaction: Through the development of APP and real-time sensing of intelligent interactive devices, the big data of activities in the "Smart Tree" is collected, and a network-based data platform and online communication forums based on cloud technology are formed.

物联网：基于智慧树大数据平台，建立对"蔬菜"这一产品的互联网关联。实现对蔬菜的种植、采摘、线上预约、售卖的一体化流程。
Internet of Things: Based on the "Smart Tree Big Data Platform", an Internet connection for the "vegetables" product is established which can achieve an integrated process for growing, picking and planting vegetables as well as online booking and selling of vegetables.

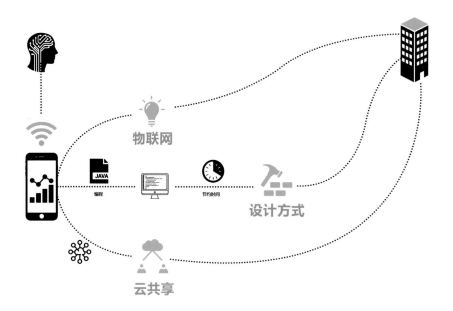

运营方式
OPERATION MODE

线下农场与线上运营相结合，完成物联网的整个过程。使用者可以在农场中心选购种子，栽培到指定区域，手机联网观察操作整个生长周期；同时加入互动论坛平台，发布成熟蔬菜，亦可线上认购某区域的成熟蔬菜，到指定区域取走。
The offline farms are combined with online operations to complete the entire process of the Internet of Things. Users can purchase seeds in the farm center, cultivate it in the specific areas, observe the entire growth cycle of the vegetables through mobile phone networking. At the same time, they can join the interactive forum platform to publish ripe vegetables, or subscribe to ripe vegetables online, and then take them away from the designated areas.

节能技术分析
ENERGY-SAVING TECHNOLOGY ANALYSIS

想象每一个居住单元是组成垂直海绵城市概念的细胞。具体说来，外墙面收集雨水，中水系统处理废水，存储在蓄水水腔，实现居住单元的自循环水系统；同时，墙面的通风系统将污浊空气经由绿植过滤为新鲜空气进入室内。每一个居住单元收集多雨水送到景观水柱，灌溉底层景观农场。
Imagine that each residential unit is a cell that forms the concept of the vertical Sponge City. Specifically, the external wall collects rain water; the middle water system filters and treats the wastewater, then stores it in the water storage chamber to realize the self-circulating water system of the residential units. At the same time, the wall ventilation system filters the polluted air into fresh air through the green plants and enters into the rooms. Each residential unit collects rainwater and sends it to the landscape water column to irrigate the landscape farm below.

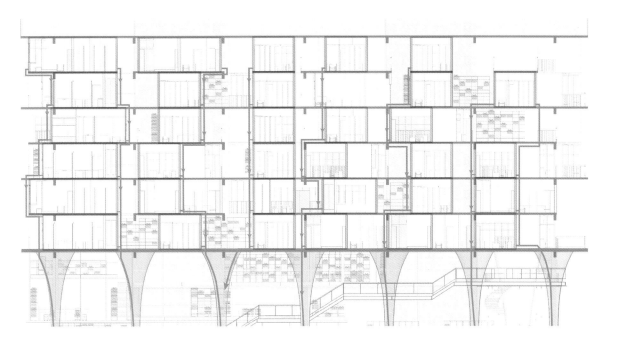

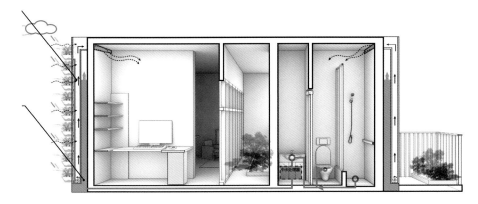

公共空间设计
PUBLIC SPACE DESIGN

将传统居住小区的配套设施分布在垂直空间中，最重要的一点是将公共空间最大化地对公众敞开，其中包括创客空间、商务会议、图书馆、展厅、绿色餐厅、共享咖啡厅、影院、农场等。其中，农场的概念贯穿了整个设计，从场地延伸至室内，并延续到整个公共空间。

在公共空间中，我们加入了灵活多变的家居设计。通过定制的模块化网格架子组成不同功能的家具，可以用于隔墙、储物、桌椅、书架等等。实现了一体化定制的家居概念。

The facilities of the traditional residential quarters are distributed in the vertical space. The most important point is to maximize the public space, including maker space, business meetings, libraries, exhibition halls, green restaurants, shared cafes, cinemas, farms, etc. Among them, the concept of the farm runs through the entire design, extending from the site to the interior and continuing into the entire public space.

In the public space, we have added a flexible home design. Furniture with different functions that are made with modulated grid or shelf can be used for custom partitions, storage, tables, chairs, bookshelves, etc. This has achieved an integrated custom home concept.

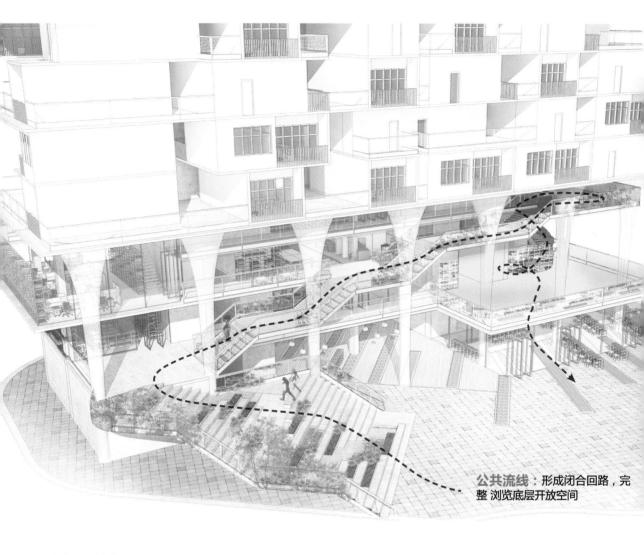

公共流线：形成闭合回路，完整 浏览底层开放空间

流线分析
STREAMLINE ANALYSIS

大台阶从广场空间抬升起来，连接各层公共空间的户外楼梯攀爬向上，到达顶层再经由与微型农场结合的旋转楼梯回到底层。提供了一个完整浏览整个公共空间的流线。整个流线交通与微型农场结合，人可以在行走时观赏并参与其中，具有观赏性和互动性。

The large steps rise from the square space, and the outdoor stairs connecting to the public spaces of each floor climb up, reaching the top floor and returning to the ground floor via a spiral staircase shared with the miniature farm. The whole streamline provides a complete view of the entire public space. The whole streamline traffic is combined with the miniature farm, and people can watch and participate in it while walking, which is ornamental and interactive.

户型组合:
Unit Combination

❶ SOHO 组团

工作模块

＋

睡眠模块

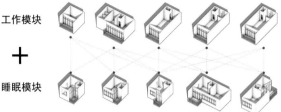

❷ 青年公寓组团

人生阶段
定制选择

单身公寓　　核心家庭

设计亮点:
Design Features

✓ **预制模块，现场装配**

卫生间模块　整体厨房模块

✓ **灵活多变的家具布置**

✓ **种植阳台 + 生活阳台**

种植阳台　　生活阳台

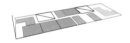

根据核心筒位置，设置避让范围
According to the location of the core tube, choose the scope of avoidance.

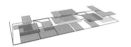

逐层生成最优平面
Generating the optimal plane by layer.

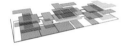

逐层生成最优平面
Generating the optimal plane by layer.

逐层生成最优平面
Generating the optimal plane by layer.

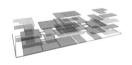

逐层生成最优平面
Generating the optimal plane by layer.

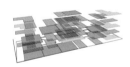

逐层生成最优平面
Generating the optimal plane by layer.

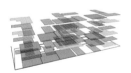

逐层生成最优平面
Generating the optimal plane by layer.

生成建筑墙体、门窗、家具等细部
Generating architectural details, such as doors, windows, walls and furniture.

最终选择 9 种典型模块组合来进行标准层平面布置，根据整数规划原理，以密度最大为目标进行生成优化过程。最终实现的标准层平面在保证密度基础上，构成灵活多变，形成丰富的公共交流空间，每层之间存在自然的差异性，最终形成具有更为丰富和复杂肌理的垂直聚落。

We finally choose 9 kinds of typical modules to carry out the layout of the standard floor generating process. The optimization process is generated according to the principle of integer programming with the goal of maximizing the density. The final implementation of the standard layer plane is flexible and variable on the basis of guaranteed density, forming a rich public communication space, and there is a natural difference between each layer, eventually forming a vertical settlement with richer and more complex texture.

荣获奖项：入围奖　　　　　　Awarded: Finalist Award

智核绿巢　　　　Intelligent Nuclear Green Nest

团队成员：董畅，赵星云，刘兵
来自：清华大学建筑设计研究院有限公司 & 北京城建集团

Team members: Dong Chang, Zhao Xingyun, Liu Bing
From: Architectural Design and Research Institute of Tsinghua University Co., Ltd.& Beijing Urban Construction Group Co., Ltd.

设计说明
DESIGN NOTESR

"尊重,利用 + 自然 + 智慧"是我们此次设计的核心理念。尊重建筑原有的结构，让室内空间与建筑，让建筑与公共空间，让公共空间与城市自然达成最大化的和谐共存，用智慧系统串联空间人与自然形成和谐高效的社区。将智核绿巢分为三个部分：微社区圈，O2O 体验店，智慧树核三部分。

"Respect, use + nature + intelligent" is the core idea of our design. Respect the original structure of the building to achieve maximum harmony and coexistence between the indoor space and architecture, between the building and public space, and between the public space and urban nature; and use the intelligent system to connect space where man and nature together form a harmonious and efficient community. The green nest is divided into three parts: the micro community circle, the O2O experience shop, and the intelligent tree core three parts.

智核绿巢——效果图
INTELLIGENT NUCLEAR GREEN NEST -- DESIGN SKETCH

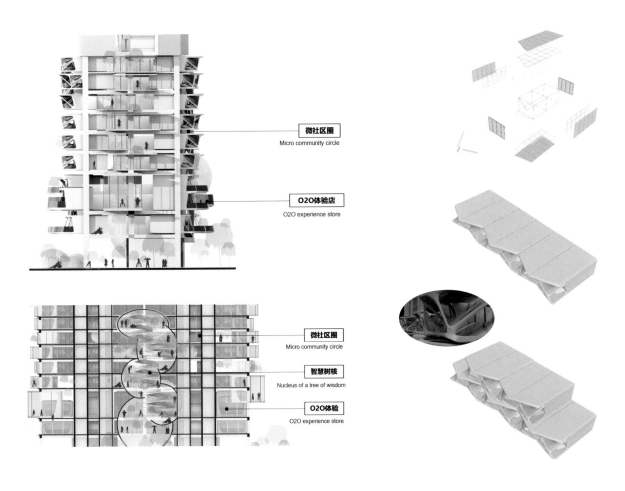

微社区圈
MICRO-COMMUNITY CIRCLE

微社区圈：打破冰冷的密闭住宅，打散成居住单元，根据个人情感的变化随意组成社群，各社群通过中心共享空间联系交流，再通过社区APP加强各社群的联系，构成线上线下人性化的微社区。

O2O体验店：建立建筑内社区商业体系，O2O无人体验店的设立实现了10分钟商圈模式，利用新技术整合了线上与线下商铺，同时在建筑内用绿化串联个体验店使之形成绿色公园式体验街，把购物与休闲健身融为一体。

智慧树核：作为整个社区的起搏器，串联人们的起居、交流、购物、运动、体验休闲，营造社区感的同时收集大数据，分析建立互联网移动端微社区，不断升级改进闭合人们的生活，服务社区的每个方面。

Micro-community circle: break up the cold closed housing; break up into residential units; according to the changes in personal emotion, form a community at will; the community communicate through the central shared space, and then through community APP to strengthen the links between the communities; and form online and offline humanized micro-community.

O2O Experience Shop: Build a community business system in the building. The establishment of the O2O Unmanned Experience Shop realizes a 10-minute business circle model. It integrates online and offline shops by using new technology. At the same time, it forms a green park-style experience street by connecting individual stores in series with greening in the building, which integrates shopping with leisure and fitness.

Intelligent Tree Core: As the pacemaker of the whole community, it connects people's living, communication, shopping, sports, leisure experience, creates a sense of community while collecting large data, analyzes and establishes the Internet mobile end micro-community, constantly upgrades and improves the closure of people's lives, and serves every aspect of the community.

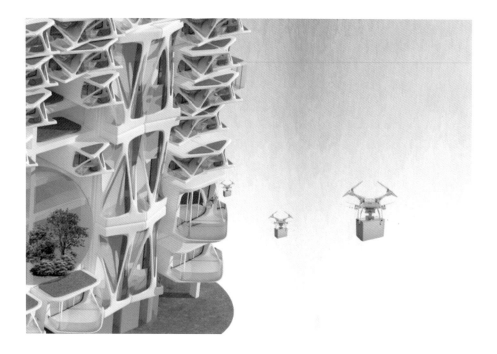

O2O 体验店——模式
O2O EXPERIENCE SHOP --MODE

最大化的增加购物的体验感，把城市森林的概念与 VR 体验店结合，用绿化空间串联各购物体验店，在中心创造公园健身体验空间来整合购物商铺，创造在森林中购物的体验感。

To maximize the experience of shopping, the concept of urban forest and VR experience shops are combined, where the green space is used to connect shopping experience shops in series. The Park fitness experience space is created in the center to integrate shopping shops and create the experience of shopping in the forest.

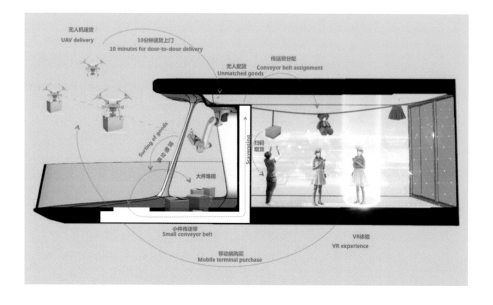

智慧树核——模式
INTELLIGENT TREE CORE--MODE

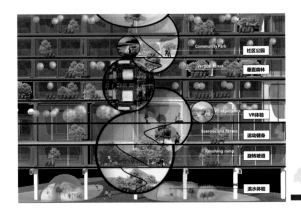
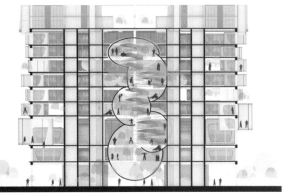

公共性
PUBLICITY

可变性
-VARIABILITY

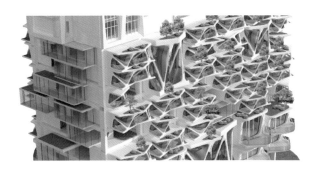

设计解析——垂直森林
DESIGN ANALYSIS--VERTICAL FOREST

设计解析——智能技术
DESIGN ANALYSIS--INTELLIGENT TECHNOLOGY

设计解析——融合之美
DESIGN ANALYSIS--THE BEAUTY OF FUSION
社区 + 绿化 + 商业 + 体验
COMMUNITY + GREENING + BUSINESS + EXPERIENCE

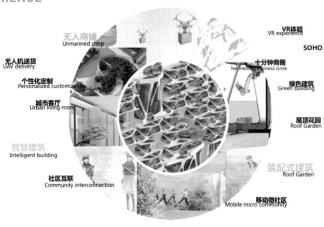

荣获奖项：入围奖　　　　　　　Awarded: Finalist Award

智能细胞系统
——未来垂直生活社区

SMARTCELL SYSTEM
—— Future Life in a Vertical Community

团队成员：Tonny LEUNG, Liu Song Song, Ren Yue, Luo Huan
来自：陆唐来建筑有限公司

Team members: Tonny LEUNG, Liu Song Song, Ren Yue, Luo Huan
From: Lu Tang Lai Architects Ltd.

建筑像树一样生长和维持。家庭作为细胞相互作用和共存。我们称每家每户都是单元细胞。由于社会的多样性和多样化的内在需求，每个单元细胞都经历了突变、生长、繁殖、缩小等过程。

Building grows and sustains as a tree. Households interact and coexist as cells. We call each household a HOUSECELL. Due to social variety and diversified internal need, each HOUSECELL undergoes mutation, growth, reproduction, downscale and so on.

一直以来，建筑都是按照开发商的意愿来设计，最终住户的愿望一直被忽视，好像"树"与"细胞"的相互关系。树的外在形态应该能反映所有细胞的协同，而不是细胞根据树的外形来适应。因此，每个单元细胞的内在独特性，都应该受到关注，使建筑物根据每个单元细胞的需要而生长。

From past to present, buildings are designed according to developers' desire. Ultimate users' wishes are always ignored. This is contradictory to Tree- and- CELL hierarchy. A tree should reflect synergy of all cells, but not vice versa. Each HOUSECELL, due to its intrinsic characteristic, should be the focus so that the building is shaped to suit each HOUSECELL's need.

在中国，每人日常的衣食住行都充斥着支付宝和淘宝等高科技媒介，技术上有可能开发一个 手机APP，或者在智能电视和电脑中开发一个交互式界面来控制和跟踪单元细胞的状况。我们在此称之为智能细胞系统。

In China, human behaviors in terms of living, travelling, buying and eating are obsessed with high-tech media, such as Alipay, Taobao. It is possible to develop an APP in mobile phone or an interactive interface in smart TV and PC to control and keep track of HOUSECELL condition. We call this SMARTCELL SYSTEM.

智能细胞系统可以用于五个不同的功能。今后，此系统可以通过更多的子系统进行优化。

SMARTCELL SYSTEM can serve as five different purposes. In future, the system can be optimized with more subsystems.

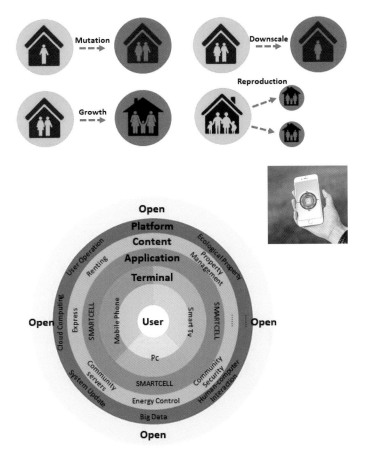

1. 模块整合
INTEGRATION OF LIVING MODULES

通过将不同的模块组合在一起，每个楼层都可以按照开放平面来设计，与其他楼层不相同。

By combining different modules together, each floor can be designed according to the open plane, which is different to other floors.

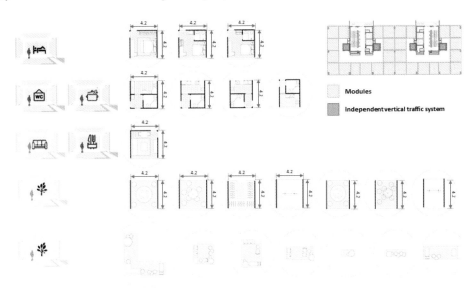

2. 智能能源控制
SMART ENERGY CONTROL

单元细胞可以在智能细胞系统中读取能耗数据记录。通过限制或控制日常活动，每个单元细胞都可以自律地帮助减少地球资源的耗散。

HOUSECELL can read energy consumption by data recording in SMARTCELL SYSTEM. By limiting/controlling daily activities, each HOUSECELL can help to reduce earth resource dissipation through shelf-awareness.

单元细胞以模块的形式出现。它们的大小反映了日常生活的能源需求量。一般而言，单元细胞面积越大，能耗越多。根据其面积大小，不同数量和规模的可再生能源设备的连接每个细胞。

HOUSECELLs exist in modules. Their sizes reflect the amount of activities taking place. In general, the larger the HOUSECELL size is, the more energy it consumes. Different number and size of renewable energy source devices are attached according to their size.

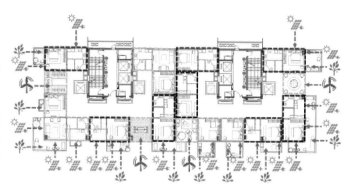

- Wind Turbine
- Solar Panel
- Plantation
- Fiber Cement Board

3. 智能社区服务
SMART COMMUNITY SERVICE

社区服务区位于低区，由常规模块组成，类似于较高楼层的单元细胞。社区服务职能的领域可以很容易地增加或减少，也可以更换或复制。用户可以使用智能细胞系统订购社区服务。智能细胞系统运用大数据收集使用过的社区服务，预测最终住户的未来需求，优化他们的日后服务。

Community services area is at lower floors and composed of regular modules, similar to HOUSECELLs at higher floors. The area of community services functions can be added or reduced easily, and also, they can be replaced or reproduced. Users can use SMARTCELL to order community services. SMARTCELLs collect the big data of the community services' usage to forecast the ultimate users' future need and optimize their services.

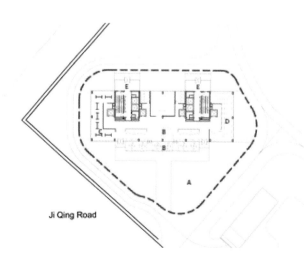

Ji Qing Road

4. 智能社区安保和物业管理
SMART COMMUNITY SECURITY AND PROPERTY MANAGEMENT

随着闭路电视连接在智能细胞系统中，所有的单元细胞都成为共享社区的看守人。这加强了小区的凝聚力，从而创造了一个和谐的环境。物业管理团队也通过智能电池系统管理每台设备。

With CCTV signage connected in SMARTCELL SYSTEM, all HOUSECELLs become vigilantes of shared community. This strengthens cohesion of all HOUSECELLs, thus creating a harmonious environment. Property management team manages every equipment by Smart Battery System as well.

5. 智能快递
SMART EXPRESS LOGISTICS

每个单元细胞都有自己的快递寄存柜，由智能细胞系统控制。系统控制柜子检查家庭是否在公寓里，什么样的快递等，以帮助快递员提供更好的服务。

Each household has his own express cabinet, which was controlled by SMARTCELL SYSTEM. System controls the cabinet to check whether the household is in the apartment, what kind of express it is and so on, in order to help the courier to provide better services.

整合了上述智能细胞系统的所有功能后，建筑外观效果和阳台特写和见下图。建筑犹如树木般生长，而随着时间流逝，单元细胞将会有机地产生变异和繁殖，生生不息。

After incorporating all elements of SMARTCELL SYSTEM, the exterior view and close-up view of balcony are shown as below. As the building grows like a tree, HOUSCELLS mutate and reproduce vertically as time goes.

今天 Today | 10 年 10 Years | 20 年 20 Years | 30 年 30 Years

荣获奖项：模块奖　　Awarded: Unit Module Prize

ADAPTIVE.COM MUNITY

团队成员：赵鹏程，李力，姚海峰，吴佳倩，陈宇龙，钱华
来自：东大-中南置地未来住区联合研究中心

Team members: Zhao Pengcheng, Li Li, Yao Haifeng, Wu Jiaqian, Chen Yulong, Qian Hua
From: Southeast University-Joint Research Center for Zhongnan Zhidi Future Residential Settlements

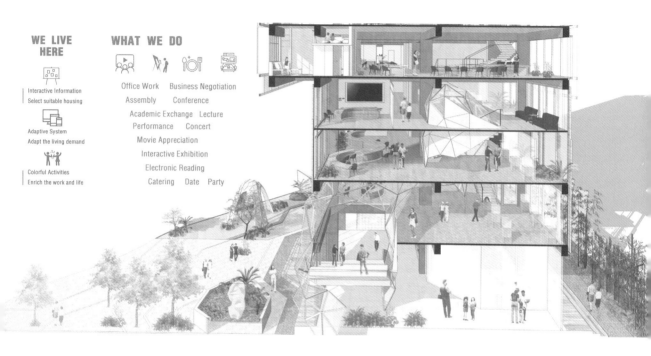

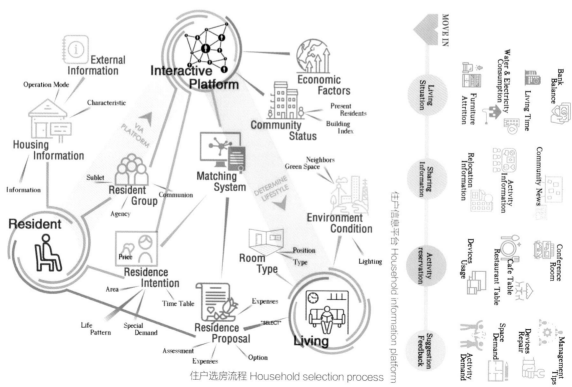

住户选房流程 Household selection process

将时间纳入设计要素，参赛者提出了自适应垂直社区的概念。结合实时数据和用户需求，使用相关算法计算当前状态下最佳的功能类型和排布方式。从而打造一个动态的、保持活力的未来青年社区。

Incorporating time into design elements, we proposed the concept of adaptive vertical community. Combining real-time data and user requirements, we used algorithms to calculate the best function types and arrangement in the current state, to create a dynamic and active community for the future youth.

在每个入住的用户输入个人相关信息之后，系统会自动匹配得到最合适的居住位置和房型。并在生活的过程中，根据房内居民的生活状态，不断调整公共空间的功能。在茧空间的连接下，整个建筑在垂直方向上保持着不同方面的联系。

After users enter personal information, the system will automatically match the information to get the most suitable living location and room type. In the further process of living, according to the living conditions of the residents in the room, the function of the public space is constantly adjusted. Under the connection of the cocoon space, the entire building stays in touch in the vertical direction in different aspects.

上半部分的居住单元可以根据用户的不同需求进行家具布置的变换，同时再单元之间可以实现单元组合的变换。

The layout of the furniture in the living unit in the upper part can be changed according to the different needs of the user. Furthermore, the unit combination can be transformed between units.

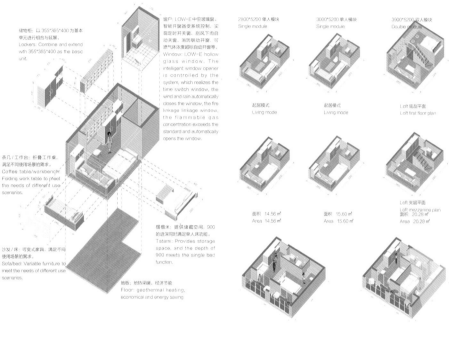

在公共空间层，每一块空间的功能随着住户的需求变化而变化，系统会根据需求和现状，计算出最优的功能调整办法。

In the public space floors, the function of each block of space can be changed based on the needs of the residents. This system calculates the optimal function adjustment method according to the needs and the status quo.

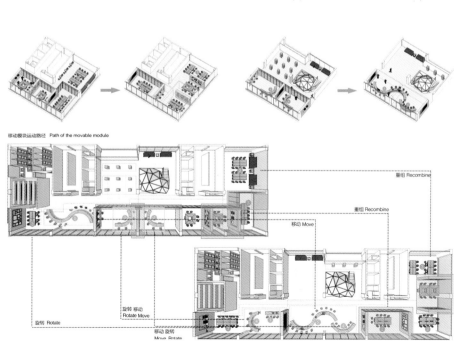

| 荣获奖项：入围奖 | Awarded: Finalist Award |

Uplift Microcity

团队成员：Keyu Xiong
来自：哈佛大学设计学院

Team members: Keyu Xiong
From: Harvard Graduate School of Design

描述
DESCRIPTION

"人与自然和谐共处（天人合一）"是中国古典哲学中的一个基本理念，其中山水田园生活是一种深深植根于传统的理想生活方式。然而，大规模的城市化发展使这种传统慢慢消散了，同时也对能源资源和基础设施造成了巨大压力，并且造成了城市生活方式和人类对生活质量的需求脱节。这个设计理念为解决城市人口密集、土地稀缺和住房危机提供了可行性方案的同时，也提出了一种新的城市愿景和可持续居住模式，以恢复人与自然之间的和谐。

"Harmony between humans and nature (tianren heyi)" is a fundamental concept in classical Chinese philosophy and living within the landscape is the ideal lifestyle that is deeply rooted tradition. However, massive urbanization has been dissipating the traditions and heritage, placing a strain on energy resources and infrastructure, and creating a huge disconnection between how we live in our cities and what we need as human beings for quality of life. The design concept presents a new urban vision and sustainable residential model to restore harmony between humans and nature in urban living, while providing a viable solution to fight for land scarcity and housing crisis in densely populated cities.

该提案将"智慧树"设想成一个自给自足的立体城市的缩影——一个集生活、工作、购物、饮食、娱乐、教育、服务、能源和食品生产于一体的共生系统。就像一个有机的共生机制，所有的项目相辅相成，互相受益。这个系统由两部分组成：上层模块化公寓和下层公共垂直社区以及各自的循环路线。

The proposal envisions "Smart Tree" as a miniature of self-sufficient vertical city – an aquaponics system integrates living, working, shopping, eating, recreation, education, service, energy, and food production as one. Like an organic symbiotic living mechanism, all programs inter-depend, support, and benefit each other. It consists two parts: an upper modular apartments and public vertical community on the lower levels with respective circulation routes.

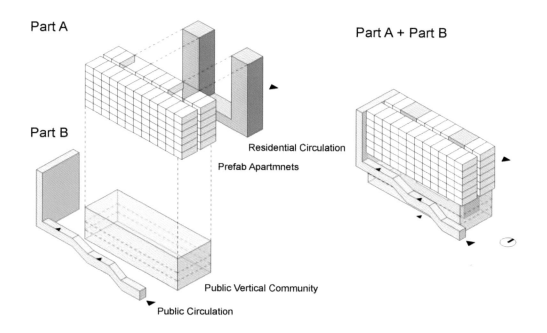

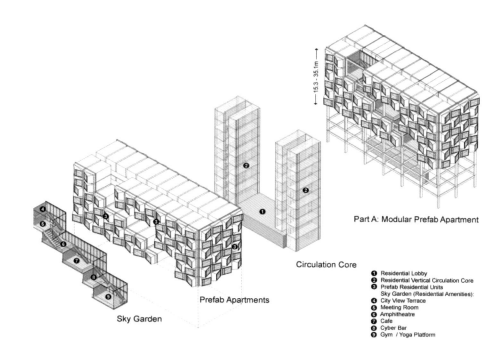

A 部分：上层模块化组合式公寓
PART A: UPPER MODULAR PREFAB APARTMENTS

上层模块化公寓（A 部分）从建筑北侧通过一楼的住宅大堂进入。由模块化组合式公寓和阶梯式空中花园组成，为居民提供各式各样的便利空间。紧凑型组合式居住单元便于运输和安装，同时又足够灵活，可以在城市的任何位置安装。模块化组件使这种居住单元具有适应性，可以随意调整以适应不同的居住容量要求。

The upper modular apartments (Part A) are accessed from the north side of the building by a residential lobby on the ground floor. It is composed of modular prefabricated apartments and a stepped sky garden with various amenity spaces for the residents. The compact prefabricated residential units are easy to be transported and installed; while at the same time, it is flexible enough to be adapted to any location across the city. Its modular subassemblies make it adaptable, allowing adjustments to be made in order to suit different residential capacity requirements.

组合式住宅单元类型
PREFABRICATED RESIDENTIAL UNIT TYPES

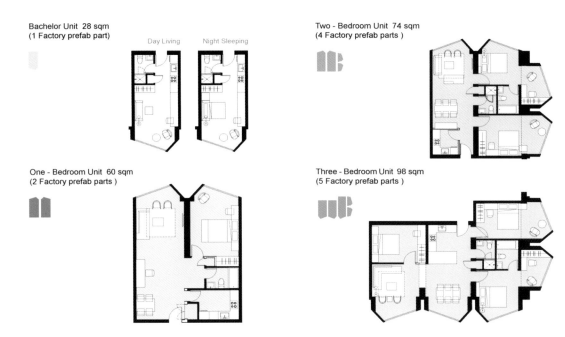

B 部分：下层公共垂直社区
PART B: LOWER PUBLIC VERTICAL COMMUNITY

在下层的公共楼层（B 部分），南立面和西立面均环绕着共生垂直农场，为超市和餐馆提供天然健康的蔬菜和鱼类，还可作为食品卫生研究与发展中心的实验室，以及可持续发展教育中心的工作间。

At the lower public levels (Part B), an aquaponics vertical farm wraps around the south and west facades, supplying natural health vegetables, and fish for supermarket and restaurant. It also services as a laboratory for Food and Health Research and Development Center, and a workshop space for Sustainability Education Center.

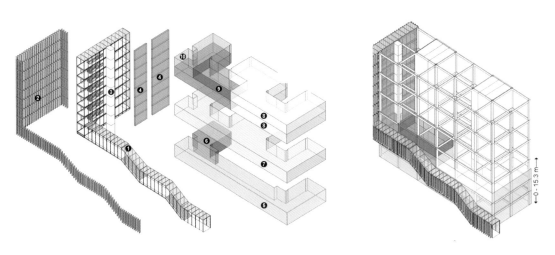

1. 公共循环绿色通道　Green Walkway for Public Circulation
2. 水耕系统散热翅片　Hydrophonic System Fins
3. B 部分垂直运输电梯　Elevator for Part B Vertical Transport
4. 鱼缸　Fish Tank
5. 有机食品超市　Organic Food Supermarket
6. 装载区　Loading Area
7. 有机食品餐厅　Organic Food Restaurant
8. 食品卫生研发中心　Food and Health R&D Center
9. 可持续发展教育中心　Sustainability Education Center
10. 共生垂直农场加工厂　Aquaponic Vertical Farm Process Plant

太阳辐射效益和能源效率
SOLAR RADIATION BENEFIT ENERGY EFFICIENCY

整个建筑以太阳能为动力,其立面开口和几何结构的设计是为了最大限度地提高太阳能辐射效益,并根据其独特的空间条件和现场定位优化能源效率。

The whole building is powered by solar energy, and its facade opening and geometry are designed to maximize solar radiation benefit and to optimize energy efficiency according to its distinctive spatial conditions and orientation on site.

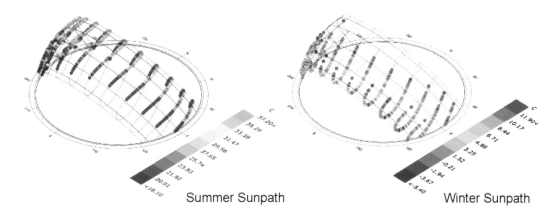

养耕共生
AQUAPONICS

养耕共生是一个将渔业和种植业在一个共生环境中结合起来的封闭的循环系统。这是一个完全自然的过程,鱼吃食后排泄废物,植物吸收这些由有益菌转化的营养辅助生长,同时净化水质,之后干净的水再次循环用于养鱼。这个过程所需水量是传统农业的六分之一,并且无需使用除草剂、杀虫剂或其他有害化学物质,保证了鱼类和植物的安全健康食用性。共生系统为超市和餐馆提供有机鱼类和农产品,同时也可作为可持续发展教育的参观中心以及食品卫生生物研发中心的实验室。

Aquaponics is a closed, recirculation system combining fishing farming and plant growing in a symbiotic environment. It is a completely natural process that fish eat food and excrete waste, which is converted by beneficial bacteria to nutrients, and plants grow by absorbing these nutrients while helping purify water, and then the clean water can be returned to fish farming. It only uses one sixth of water that traditional farming requires. No herbicides, pesticides or other harsh chemicals can be used, making the fish and plants healthful and safe to eat. The aquaponic system supplies organic fish and plants for supermarket and restaurant, and also as a visiting centre for sustainability education, and a laboratory for Biological Research and Development Center for Food and Health.

- 沉淀池 & 生物滤池:沉淀颗粒物,捕捉剩下的食物和分离的菌膜,培养硝化细菌,将氨转化为硝酸盐,并将富硝盐水抽回水耕系统
- Settling basin & Bio filter: settling out particulates, catching uneaten food and detached biofilms, culturing nitrification bacteria, converting ammonia into nitrates, and pumping nitrate-rich water back into hydrophonic system

- 水耕系统:植物将硝酸盐作为食物吸收,同时可以清洁水质(可旋转的水耕散热翅片可以旋转到不同的角度以适应阳光,并且能够快速翻转到内部以便于冬季采收和人工照明)
- Hydrophonic system: plants absorb the nitrates as food while cleaning the water (rotatable Hydrophonic Fins can be rotated different angles to adapt to sun light, and flipped to inside for easy harvest and artificial light in winter)

- 污水坑:充气,干净水循环至鱼缸
- Sump: aerated, clean water is re-circulated back to fish tanks

- 集成太阳能电池:为农场运营供电
- Integrated solar cells: powering up the operation of the farm

- 屋顶雨水收集
- Rainwater catchment on roof

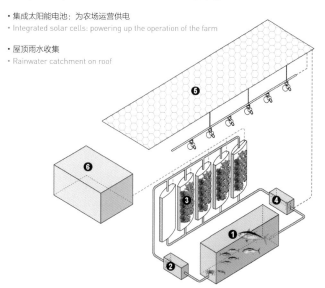

| 荣获奖项：入围奖 | Awarded: Finalist Award |

垂直森林   Vertical Forest

团队成员：Benny Lee，Chiu Ming，Paul Mui，Agnes Hung
来自：Bread studio

Team members: Benny Lee, Chiu Ming, Paul Mui, Agnes Hung
From: Bread studio

概念
CONCEPT

1. 树木可以遮阴，湖泊有助于降温
2. 树冠可以遮阴，水池有助于降温
3. 多层次的树木和水池
4. 只在地面上的花园
5. 花园延伸到高层

1. Tree provides shade; lake promotes cooling
2. Tree like canopy provides shade; pool promotes cooling
3. Multiple levels of trees & pools
4. Garden on ground level only
5. Garden extended up to high level

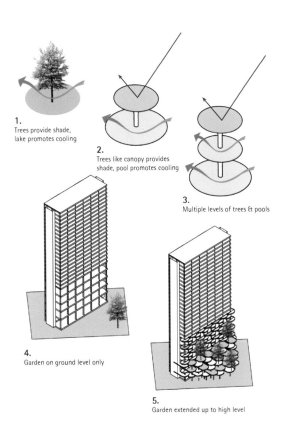

底层平面图 1:200
GROUND FLOOR PLAN 1:200

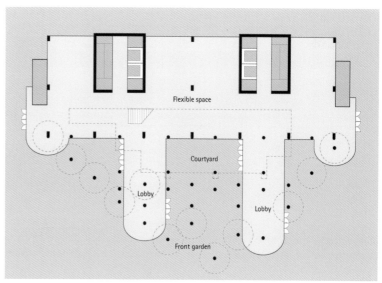

Site Plan 1:1000

一层平面图 1:200
FLOOR PLAN LEVEL 1 1:200

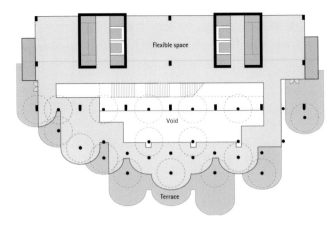

二层平面图 1:200
FLOOR PLANS – LEVEL 2 1:200

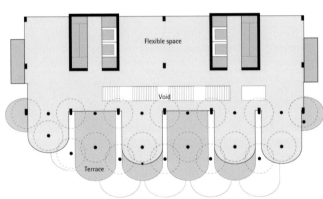

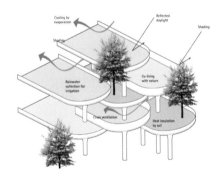

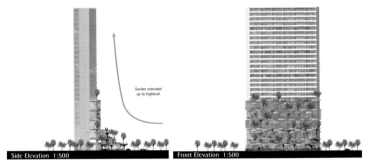

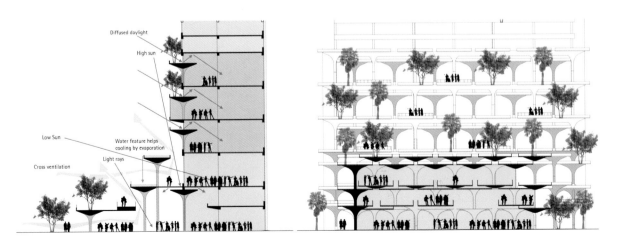

徐卫国
Xu Weiguo

清华大学建筑学院教授、博士生导师、建筑系主任
中国建筑学会数字建造委员会副主任

Professor of Architecture School in Tsinghua University, Doctoral Tutor, Director of the Department of Architecture, Deputy Director of the Digital Construction Committee of the Architectural Society of China.

构筑数字建筑的产业网链
Constructing the Industrial Network Chain of Digital Buildings

数字建筑历史不长，但发展很快，目前已经建立起基本的理论框架和体系，而随着未来智能建造的实现，"数字建筑产业网链"也正在形成之中。

Although not enjoying a long history, digital buildings have developed very quickly, and have established a basic framework and system of theory at present. Along with the realization of future intelligent construction, "the Industrial Network Chain of Digital Buildings" has been in the formation.

数字技术并不是为了生成酷炫造型，仅仅追求形式，它依然为建筑的本质服务，其核心是如何通过数字技术让建筑设计满足人的需求。从设计角度而言，数字技术探讨如何实现人文、环境、建筑三者的最优关系。因此，其中一方面是对数字设计理论的研究，比如数字图解的设计方法、设计思想、建构概念、算法生形等等。最近我出了一本书《生物形态的建筑数字图解》，展现的就是这些工具性、技术性的研究结果。书中介绍了30个来自于生物形态的生形算法，这种算法能够帮助建筑师有效地实现形式生成的基本关系。

Digital technology is not to create cool shapes or only to pursue modality. Instead, it still serves for the essence of building, and its core consists in how to make building design meet people's needs through digital technology. From the perspective of design, digital technology discusses how to realize the optimum relations among humanity, environment and building. Therefore, we conduct research on the theory of digital design among others, such as the design method, design ideas, construction concepts and algorithm generating, etc. of digital diagrams. I have recently published a book *The Digital Diagram of Buildings from Bio-forms*, which just presents these instrumental and technical research findings. This book introduces 30 generating algorithms sourced from bio-forms, and such algorithms could help architects effectively realize the basic relations of form generating.

目前，我们对"智慧建筑"的定义主要局限在建筑设备系统及屋宇自动化管理等方面，但从建筑的全生命周期看，设备系统及屋宇管理仅是时间序列上靠后的两个环节，其他包括建筑策划、建筑设计、构件加工、施工建造等，尚没有纳入"智慧建筑"的范畴，因此，必须对智慧建筑的既有定义进行拓展。

At present, our definition of "intelligent building" is mainly limited to building equipment system and automatic management on buildings. But as seen from the full life cycle of buildings, equipment system and building management are just two sections which lay back in time series, and others like building planning, building design, component processing, and construction, etc. have not been taken into the scope of "intelligent building". Therefore, it is necessary to expand the existing definition of intelligent building.

"智慧建筑"应当指房屋建筑的全过程及各专业,充分利用数字技术来实现建造目标。在包括设计阶段、构件加工阶段、施工阶段、物业管理阶段等房屋建筑的"全过程"中,建筑、结构、水暖电、施工以及材料配送、构件加工、施工机械、物业管理等"各专业"共同参与,共享数字流。从建筑方案开始,经过后续阶段及各专业不断添加、修改、反馈、优化形成建筑信息的数字化集成。以数字流为依据,借助互联网及物联网、CNC 数控设备、3D 打印、机器臂、AR&VR、人工智能等技术及机械,未来,房屋建造将实现一种"数字建筑的产业网链",从而实现高精度、高效率、环保性的建造与运营服务。

"Intelligent building" should refer to the full use of digital technology to realize construction objectives in the full process and each specialty of house building. In the "full process" of house building, including designing stage, component processing stage, construction stage, and property management stage, etc., "all specialties" like building, structure, HVAC, construction and material distribution, component processing, construction machinery, and property management, etc. jointly participate in the building and share digital stream. The digitalized integration of building information is formed by starting with the building scheme, going through the follow-up stages, and through the constant adding, modification, feedback and optimization of each specialty. Based on digital stream, and in virtue of the technologies and machinery such as Internet and Internet of Things, CNC equipment, 3D printing, robotic arm, AR&VR, and artificial intelligence, etc., in future, housing construction will realize an "industrial network chain of digital building", thus realizing high-precision, high-efficiency, and environmental-protection construction as well as operation services.

今天,设计工具和建造工具的新发展有力地支持了这一全产业链的形成。在数字设计工具方面,如今的软件和程序开发,已使得建筑师能够很容易使用这些工具进行建筑设计,甚至无需建筑师自身掌握编程语言,即可应用软件进行数字建筑的设计。近些年以来,建造方法也有所进展。3D 打印技术也被运用到建筑中。这为我们实现建造一些复杂形体提供了可能,而学校、实验室对此的研究结果更为实际建造、实际项目提供了基础。比如我们在 3D 打印混凝土方面做了深度研究,一开始是为了实现一些不规则的形体,因为没法建造,所以自己开发工具,最终产生了 3D 打印混凝土建造系统。我们把研究还延伸到了用机器臂 3D 打印砂浆及砌筑砖墙一体化进行建造的开发,这样产生了机器臂自动砌筑系统,可以自动砌筑不规则墙体。

Today, the new development of design tools and construction tools strongly supports the formation of this full industry chain. In terms of digital design tools, today's software and program development has enabled architects to use these tools to carry out building design easily, and even to use software to design digital buildings without mastering the programming languages. In recent years, construction methods have also been improved. 3D printing technology has been applied to buildings. This makes it possible for us to realize the construction of some complicated shapes. The results of schools and laboratories' research on this aspect have even provided a foundation for practical construction and practical projects. For example, by carrying out deep-going research on 3D printing concrete initially for realizing some irregular shapes, which cannot be constructed, we have developed tools by ourselves, and finally, produced the 3D printing concrete construction system. Also, we have extended our research to the construction development with the integration of 3D printing of mortar and brick-wall building with robotic arm, thus producing the robotic arm automatic building system, which can realize the automatic building of irregular walls.

未来,建筑工业的升级和发展目标是数字建造和智能建造。智能建造不光是建造,更要求设计和建造相联系,因为两者之间具备产业链关系,没有数字设计和数字软件,就不可能有加工和建造。而从整个领域来说,数控、制造和建筑设计已结合在一起。建筑设计能够借助于工业制造、产品制造工具和设备来进行建筑构建,甚至完成更大体量建筑的直接建造,由此实现行业的重要突破。

In the future, the upgrading and the development goals of building industry are digital construction and intelligent construction. Intelligent construction does not only means construction, but requires the connection of design and construction, since the two factors have an relationship of industry chain. Without digital design and digital software, there will be no processing and construction. As in the aspect of the whole field, numerical control, manufacturing and building design have been combined together. Building design can carry out building construction and even realize the direct construction of larger buildings in virtue of industrial manufacturing, product manufacturing tools and equipment, thus realizing important breakthroughs of the industry.

Henriette H. Bier

荷兰代尔特理工大学建筑与环境建造学院 Robotics Building 副教授
德国安哈尔特应用科学大学德绍建筑学院客座教授

ssociate professor of Robotic Building, Faculty of Architecture & Built A
A Environment, TU Delft
Guest Professor Robotics, Dessau Institute of Architecture, The Anhalt
University of Applied Sciences, Germany

机器人建筑时代
Architecture in the Age of Robotic Building

人类与机器如何合作，为建筑设计和施工带来改变，是我们这几年来一直关注的问题。如今，建筑机器人正越来越多用于建筑组件的生产、组装和操作上，由此帮助建筑业转型升级。未来，机器人建造技术将变革下一代建筑的生产和施工方式，让我们进入一个机器人建筑时代。

The question that how integration of humans and robots changes the design and construction of buildings has been focus of our research for several years. Nowadays, robots speed up production and assembly and operation of building components, contributing to the transformation and upgrading of the building industry. In the future, the robotic technology will change the production and construction methods of the next architecture, allowing us to enter an era of robotic building.

机器人是未来建筑施工的核心科技。实际上，行业内已经开始在这方面进行试验。我们期待，未来整个建筑过程都能实施自动化。专业的建筑机器人正在逐步从研发走向落地。不同类型的机器人将在建造与建筑运营的不同阶段发挥作用，我们在这一方面的探索已经超过十年。

Robots are the core technology for building construction in the future. In fact, the industry has already been experimenting with this technology, and we expect that in the future the whole building process is going to be automated. Construction robots are gradually moving from research and development to playing increasingly an active role in the construction of buildings. Various robots are going to play a role in the different phases of the construction process, and also in the operation of buildings, which we have been already exploring for more than a decade.

最终，我们的目标是将机器人融入建筑施工过程和建筑环境中，这将是建筑工业的一次彻底变革。通过计算机设计、机器人生产、自动化操作等技术，整个建筑生产都将变得更高效、更简单。此外，包括智能温度控制等自动环境监控技术，以及根据个体调节空间配置等一系列技术，将为创造更具适应性与环保性的下一代建筑提供条件。

Ultimately our goal is to integrate robots into the building processes and the built environment, which will imply a radical change of the building industry. Through computational design, robotic production and automatic, the entire building process will become more efficient and effective. Furthermore, automated environmental control involving sensor-actuators for intelligent temperature, ventilation, etc. control, and individual adjustment of spatial configuration will provide conditions for creating more adaptive and environmentally friendly next architecture.

自动化的建筑施工与运营，并不排除人类的参与，相反，其更依赖于人机互动与人机协作，从而保证环境、人员和建筑物之间的良好互动。由于涉及信息与物理两种界面的交互操作，如何优化人机协作的方式，将是机器人建筑时代的一大挑战。

Automated building construction and operation does not exclude human participation. Rather, it relies on human-computer interaction and human-robot collaboration, which both ensure communication and exchange among environment, people and buildings. Since it involves the interactive operation in both information and physics, how to optimization this communication and exchange will be the challenge that needs to be addressed in the era of robotic building.

当前整个建筑界，机器人建筑技术的应用才刚刚兴起，在目前的技术水平下，机器人技术可以被应用在建造生产、组装和材料处理上。此外，机器人技术和人工智能还被应用于室内气候与空间调节中。未来，我们还需要进一步探索这些新科技的潜力，为建筑服务。世界各地的建筑学院，都开始研究并设法解决这一课题，教授学生怎样利用这些科技。然而，不仅在中国，全球范围从事这项工作的研发团体依然偏少，因此，我们需要更大的群体，发挥更多的努力来参与这个主题。

At present, the application of robotic technology has just emerged in the construction industry. At the current technical level, robotic technology can be applied in the production, assembly and material handling processes. Furthermore, robotic technology and artificial intelligence are applied for indoor climate and spatial regulation. These technologies need further exploration of their potential for architecture. Architecture schools around the world begin to explore and teach students to work with these technologies. However, the R&D groups that do this work around the world are still underrepresented, so we need a larger effort and bigger team to focus on this topic.

作为下一代建筑大奖的评审，我期望看到更多运用机器人参与建造的项目。大赛中，一些参赛者提出的方案也恰好囊括了这部分理念，因此，本次大赛的举办十分富有意义。我期待看到建筑行业采纳这些未来建筑的理念。我相信，建筑机器人将以积极的姿态加速发展，不仅促进建筑与建筑工程的革新，而且在解决资源稀缺、气候变化与人口过剩等社会问题上发挥积极作用，在全球实现愈加广泛的应用。

As a judge of "The Next Architecture" Award, I look forward to seeing the use of robots in buildings and building processes. Some of the participants have been making proposals that include some of those ideas, so this is very promising and I look forward to seeing these concepts of future architecture being adopted by the construction industry. In the future, building robotics will contribute to not only improving buildings and building processes but also addressing societal challenges such as material scarcity, climate change, overpopulation, etc., and realizing more and wider architectural applications in countries around the world.

张雷
Zhang Lei

江苏省设计大师，南京大学建筑与城市规划学院教授
可持续乡土建筑研究中心主任
张雷联合建筑事务所创始人兼总建筑师

Design Master of Jiangsu Province, professor of School of Architecture and Urban Planning, Nanjing University. Director of Sustainable Local Architecture Research Center. The founder and chief architect of AZL Architects.

生态科技 回归自然
Eco-technology, Returning to the Nature

下一代建筑的发展趋势可归结为三点：即反映新的价值观、新的审美及结合技术革命的发展。这三方面将是推动下一代建筑发展的最大作用力。首先，下一代建筑新的价值观，是建筑回归自然。回归自然是人类发展到一定程度的必然需求，如何让人与自然更好地相处，也是建筑师需要首要考虑的问题。下一代建筑，必然会因这种对自然的需求而更具生态性。第二，对比当前建筑的审美，下一代建筑必将更具地域性特点。未来建筑会进一步结合地域文化、地域气候及地域特征，也因此产生出对建筑更好的价值审美。

The development trend of The Next Architecture could be summarized into three points: that is to say, reflecting new values, new aesthetics, and the development combining the technological revolution. The three aspects will be the greatest driving forces which promote the development of The Next Architecture. Firstly, the new value of The Next Architecture is that buildings should return to the nature itself. Back-to-nature is an inevitable demand of the human beings after developing to a certain degree. Therefore, how to make the human beings get along better with the nature is a problem that architects shall put in the first place. The Next Architecture will consequentially be more ecological due to such demand on the nature. Secondly, compared with the aesthetic of the present buildings, The Next Architecture will certainly have more regional characteristics. Future buildings will be further combined with regional culture, regional climate and regional characteristics, thus producing a better aesthetic value for buildings.

最后，需要科技进步来驱动下一代建筑的发展，与未来建筑结合的科技包括人工智能及生态技术的发展、新材料的运用、新结构系统的推广等。当前，科技正以无与伦比的速度向前发展。相对而言，建筑行业还未能同步前进。高科技与新建筑的结合，是下一代建筑重要的议题。人工智能将会带来从设计、建造、控制到使用的建筑全流程改变。3D打印技术很可能改变未来建筑的建造方式，让建筑生产更有效率。突破性的新材料可能更加轻巧，但受力与力学性能更好，为建筑结构提供更多样的可能性。

Finally, the development of The Next Architecture needs to be driven by technological advances. The technologies combined with future buildings involve the development of artificial intelligence and ecological technologies, the application of new materials, and the promotion of new structure systems, etc. At present, technology is developing at an unprecedented pace. Relatively speaking, building industry has not realized synchronous progress yet. The combination of high technologies with new buildings is an important issue for The Next Architecture. Artificial intelligence will contribute to the changes to the full process of buildings from design, construction and control to use. 3D-printing technology will probably alter the construction method of future buildings, and make building production more efficient. Ground-breaking new materials may be lighter, but have better stress-bearing and mechanical properties, which could provide more diversified possibilities for building structure.

在技术飞速发展的同时，科技运用依然需要为特定的价值观和目的服务。建筑需要体现人的需求。技术并不一定越高越好，它不是建筑美化的工具，也并非建筑增值的手段，而是需要反映新的建筑思想，使技术更好地为功能服务。同时，当新科技普及在更大量的建筑上时，普通人也能体验由新技术带来的更美好的生活，这是技术对于建筑积极的意义所在。

While technology develops rapidly, the application of technology still needs to match certain values and purposes. Buildings shall embody people's needs. The more advanced technology does not necessarily mean being better. It is not a tool for the beautification of buildings, nor a means for the increment of buildings. In contrast, technology ought to reflect new building ideas, and to serve functions still better. Meanwhile, when new technologies are commonly applied to more buildings, ordinary people will be able to experience a better life brought about by the new technologies. This is the positive significance of technologies for buildings.

未来，建筑最重要的价值观之一，将是推崇生态、健康与回归自然的可持续性发展。因为无论在精神上，还是在空间感受上，只有好的品位与价值观，才能帮助我们体验更理想的生活。最近几年，我们在乡村做了很多实践工作，乡村复兴是一种很好地回归自然的方式。在乡村做建筑，会更注重人、建筑、环境等和谐相处的关系，也更注重把一些环保材料、地域性材料及地域性的手工艺运用在当地建筑上，从而推动地域性传统文化的发展。这些当代美学手法，也十分适合新一代年轻人的审美需求。第一代建筑大师常说"少即是多"，很多年过去了，即便技术已不同往昔，回归自然、简约纯净的审美依然能回应人类最复杂的需求。技术和审美情感始终交织在两条线上，下一代建筑，既需要先进科技的有力推动，也需要时刻关照对自然向往的情感需求。

In the future, one of the most important values of buildings will focus on advocating the sustainable development which is ecological, healthy, and back-to-nature. No matter whether in terms of spirit or space perception, only good tastes and values could help us experience a better life. In recent years, we have done a lot of work in rural areas. The revival of rural areas is a good way to go back to the nature. To construct buildings in rural areas, we should pay more attention to the harmonious relationship among humans, buildings, and environment, etc., and also place more emphasis on applying some environmental-protection materials, regional materials and regional handicrafts to local buildings, thus promoting the development of regional traditional culture. These contemporary aesthetic techniques are very suitable for the aesthetic needs of the new-generation young people. The first-generation master architects often said "less is more". Now, though many years have passed, even though technology has changed, the aesthetic of being back-to-nature, simple and pure could still respond to the most complex needs of human beings. Technology and aesthetic emotions are always intertwined in both lines. The Next Architecture not only need to be vigorously promoted by advanced technology, but also need to always care about the emotional need of the yearning for the nature.

在此次竞赛中，我看到了年轻人对未来生活的想法与理念。他们不单把自己当建筑师和设计师，也把自己当成一个未来的使用者。反映在他们作品中的，是他们对未来生活的期望和对未来空间的想象。我们并不奢望通过一个设计竞赛，解决如此复杂宏大的社会问题。但很多方案都能在某一些方面做出比较深入的探讨，很多年轻建筑师和设计师关注可持续发展的议题，体现了本次竞赛重要的价值。

In this contest, I have seen young people's ideas and concepts about the future life. They regard themselves not only as architects and designers, but also as future users. What have reflected in their works are their expectations for the future life and their imagination about the future space. We do not extravagantly hope to solve so complicated and grand social problems through a design contest. However, many schemes have carried out relatively thought-invoking and profound discussions on some aspects. Meanwhile, many young architects and designers have paid attention to the issue of sustainable development, embodying the important value of this contest.

回顾和鸣谢

Review and Acknowledgement

回顾和鸣谢 Review and Acknowledgement

2018年,国际绿色建筑联盟、东南大学、南京长江都市建筑设计股份有限公司、中建八局第三建设有限公司、南京国际健康城开发建设有限公司、雅伦格文化艺术基金会联合主办的"智慧树——垂直社区的未来生活"国际设计竞赛得到了很多单位和个人的支持,在此以回顾,并附上致谢单位名录。

2018年4月28日,南京江北新区未来绿色智慧型建筑"下一代建筑"国际研讨会在江苏省南京市召开。会议邀请了众多院士、大师、教授和华为、西门子、霍尼韦尔、科大讯飞、广联达、中建科技、汉能、和能人居等十余家智慧智能领域的先进企业共同研讨下一代建筑。对什么是下一代建筑,富有了更为深刻和广泛的内涵。

The "Smart Tree- Future Life in a Vertical Community" International Architectural Design Competition was co-organized by the International Green Building Alliance, Southeast University, Nanjing Yangtze River Urban Architectural Design Co., LTD, The Third Construction Co., Ltd of China Construction Eighth Engineering Bureau, Nanjing International Healthcare Area Development & Construction Co., LTD. and Fondazione EMGdotART in 2018, and has received support from many organizations and individuals. The process has been reviewed here and a list of those expressing thanks is attached.

On April 28, 2018, "The Next Architecture" the International Symposium of Futuristic Intelligent Green Buildings was held in the Jiangbei New Area of Nanjing, Jiangsu Province. The conference invited many academicians, masters, professors and more than ten advanced enterprises in high-tech fields such as Huawei, Siemens, Honeywell, IFLYTEK CO.,LTD., Glodon, China Construction Science & Technology Group, Hanergy and Heneng home to discuss the 'Next Architecture'. What is the 'Next Architecture'? And we are hoping to infuse it with a richer and more profound meaning.

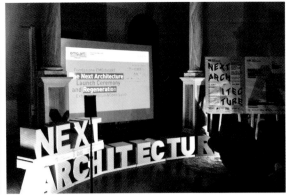

2018年5月27日，竞赛在第十六届威尼斯国际建筑双年展期间正式发布，面向全球建筑设计师征集创意方案。

On May 27, 2018, the International Competition Project was officially launched during the 16th Venice International Biennale of Architecture, to collect creative ideas from all over the world.

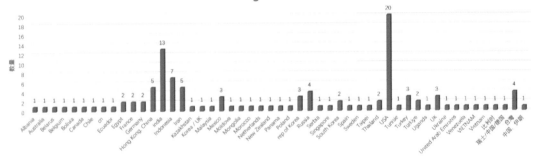

国际报名分布
International registration distribution

竞赛从2018年6月中旬开始征集，截至9月1日收集到来自全球的近百份作品，经过9月份第一轮的线上评审，最终有30份作品进入10月13日的终评，共同角逐大奖。

The open call started from mid-June 2018, and nearly 100 works from all over the world were collected by the 1st of September. After the first round of online review on September 30th, 30 works finally entered into the final round on October 13th and competed for the grand prize.

中国工程院院士、本次大赛评委会主席孟建民先生宣读了大赛评审决议书，经过终评委员会的评选和一致认证，评出一等奖1组，二等奖2组，模块奖3组，入围奖20组，同时，IAAC和雅伦格文化艺术基金会确定奖学金奖1名，共计27个奖项55万奖金和近2万欧元的奖学金。

Mr. Meng Jianmin, the Chairman of the jury committee, Academician of the Chinese Academy of Engineering, read out the the Review Resolution of the Competition. After the final jury committee's selection and unanimous certification, the First Prize was awarded to 1 place, the Second Prize was given 2 places, the Unit Module Award was given 3 places, and the finalists Award was a total of 20 places. At the same time, IAAC and Fondazione EMGdotART confirmed one scholarship award. The competition totaled an amount of 27 awards of 550,000 bonuses and nearly 20,000 Euros in scholarships.

11月20日，大赛颁奖活动在第十一届江苏省绿色建筑发展大会上举行，中国工程院院士、国际绿色建筑联盟主席、江苏省建筑科学研究院有限公司董事长缪昌文，江苏省住房和城乡建设厅副厅长刘大威，威尼斯建筑大学前校长、雅伦格文化艺术基金会主席Marino Folin等出席活动并颁奖。雅伦格文化艺术基金会主席Marino Folin为下一代建筑奖学金奖获得者颁奖。

On November 20th, the Awards Ceremony was held at the 11th Jiangsu Green Building Development Conference. Miao Changwen, Academician of the Chinese Academy of Engineering, Chairman of the International Green Building Alliance, Chairman of the Jiangsu Academy of Building Science Research Institute, Liu Dawei, Deputy Director of the Construction Department, Housing, Urban and Rural Areas of Jiangsu province, Marino Folin, President of Fondazione EMGdotART, Former President of the University of Venice Architecture, attended the Ceremony. Mr. Marino Folin, chairman of Fondazione EMGdotART, presented the awards to the winners of 'The Next Architecture' Scholarship Award.

鸣谢
Acknowledgements

国际绿色建筑联盟
International Green Building Alliance

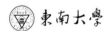

东南大学
Southeast University

南京长江都市建筑设计有限公司
Nanjing Yangtze River Urban Architectural Design CO.,LTD.

中建八局第三建设有限公司
The Third Construction Co., Ltd of China Construction Eighth Engineering Bureau

Nanjing International Healthcare Area Development & Construction Co., LTD.
南京国际健康城开发建设有限公司

雅伦格文化艺术基金会
Fondazione EMGdotART

图书在版编目（CIP）数据

智慧树——垂直社区的未来生活　建筑设计国际竞赛获奖作品集/绿色智慧建筑（新一代房屋）课题组编．—北京：中国建筑工业出版社，2019.11
（江苏省绿色智慧建筑（新一代房屋）系列丛书）
ISBN 978-7-112-24459-1

Ⅰ.①智… Ⅱ.①绿… Ⅲ.①生态建筑-建筑设计-作品集-世界　Ⅳ.①TU206

中国版本图书馆CIP数据核字（2019）第235018号

责任编辑：封　毅　张瀛天
责任校对：李欣慰

【江苏省绿色智慧建筑（新一代房屋）】系列丛书

智慧树 —— 垂直社区的未来生活
建筑设计国际竞赛获奖作品集
绿色智慧建筑（新一代房屋）课题组编
*
中国建筑工业出版社出版、发行（北京海淀三里河路9号）
各地新华书店、建筑书店经销
北京建筑工业印刷厂制版
北京富诚彩色印刷有限公司印刷
*
开本：787×1092毫米　1/16　印张：11　字数：423千字
2019年11月第一版　2019年11月第一次印刷
定价：**120.00**元
ISBN 978-7-112-24459-1
　　（34945）

版权所有　翻印必究
如有印装质量问题，可寄本社退换
（邮政编码 100037）